NORTHAMPTONSHIRE STONE

Dr Diana Sutherland graduated in geology at London University, and her research on volcanic rocks of Eastern Uganda continued as part of a team in the Geology Department at Leicester University, studying East African rocks. She edited the book, *Igneous Rocks of the British Isles*, and in 1992 became part-time Curator of igneous petrology in the Department at Leicester, until retiring in 2001. She is now an Honorary Visiting Fellow.

She has lived in Northamptonshire since 1963, and for many years was a part-time tutor in Adult Education at Leicester's University Centre in Northampton, developing a special interest in the geology and building stones of Northamptonshire. Local research has included the petrological study of Brixworth Church, and an enquiry into the geological cause of radon in Northamptonshire. She is concerned with Regionally Important Geological Sites (RIGS) in the county.

A Fellow of the Geological Society, she was awarded the R.H. Worth Prize in 1997. She and her husband have a daughter and a son.

Following page
St. Peter's Church, Northampton – according to Pevsner 'the most interesting Norman church in Northamptonshire' – was built in the twelfth century, the tower repositioned probably in the seventeenth century. The brown sandstone and ironstone are local Northampton Sand, with decorative features in stone from the Blisworth Limestone Formation. Below the tower is the prominent grave of William Smith, 'Father of English Geology', who died nearby at Hazelrigg House, home of the historian, George Baker, in 1839.

Northamptonshire Stone

D.S. SUTHERLAND

'And no County in England affording a greater Variety of Quarry-Stone
than this, or exceeding this in the Goodness and Plenty of it, upon
that account it deserves a more particular consideration.'
JOHN MORTON
The Natural History of Northamptonshire, 1712

THE DOVECOTE PRESS

The eighteenth-century garden gateway at Delapré Abbey, Northampton, made of Helmdon Stone, a limestone composed of fragments of oyster shell, from the Taynton Limestone Formation in south-west Northamptonshire.

Northamptonshire County Council

The author and publishers would like to acknowledge the assistance of Northamptonshire County Council (Built and Natural Environment), whose support has made possible the use of colour throughout this book.

First published in 2003 by The Dovecote Press Ltd
Stanbridge, Wimborne, Dorset BH21 4JD

ISBN 1 904349 17 X

© D.S. Sutherland 2003

D.S. Sutherland has asserted her rights under the Copyright, Designs and Patent Act 1988 to be identified as author of this work

Geological maps reproduced by permission of
The British Geological Survey
© NERC. *All rights reserved* IPR/37-3C

Designed by The Dovecote Press Ltd

Typeset in Monotype Sabon
Printed and bound by KHL Printing in Singapore

A CIP catalogue record for this book is available
from the British Library

Contents

Acknowledgements

I am greatly indebted to the following: Northamptonshire County Council (Built and Natural Environment) and Mrs Ann Bond (now with English Heritage); Susan Freebrey (now with The National Trust), who skilfully adapted the geological maps from the Council's digital database; Northamptonshire Libraries and Information Services for their help, and providing the illustrations (2.5, 5.7, 5.10, 6.2, 7.1); Northampton Museums for photographs (2.7, 9.12); the Northamptonshire Record Office; Bruce Bailey, archivist, and the Althorp Estate, for access to information about the Harlestone quarries; and Bruce for visits to Drayton House and Stoke Park; the owners of stone houses, for their courtesy in allowing me to examine the masonry and take photographs; English Heritage, The National Trust, for permission to photograph Canons Ashby (3.8, 3.9); slater David Ellis for information, hospitality, and providing photographs (8.4, 8.6), and stonemasons Peter Dunn and Philip McCrone for giving their time; the late Jack Stock for showing me Raunds; Dr David Parsons and the Brixworth Archaeological Research Committee; the Reverend A.J.Watkins, the Friends of All Saints', Mandy Dawkins of the Brixworth History Society; David Kench (Eydon); the former class members of the Leicester University Centre, Northampton, including Brian Statham, for their contribution to the project; staff of the British Geological Survey, particularly Dr Mark Barron for advice; Rod Branson, of Leicester University Geology Department, for assistance with photographing rock specimens; Dr Roy Clements, in the Department, for advice on the geology in Chapter 2; and Professor John Hudson, who for many years collaborated in this research, for his constructive appraisal of the manuscript. The inevitable shortcomings are my own.

My grateful thanks to all the above; and especially to my husband, John Milne, for his support in everything.

The Arms of the Masons' Company, on the early eighteenth century tomb of master-mason John Wigson at Eydon.

FORMATION	BUILDING STONES	ROCK-TYPE
BLISWORTH LIMESTONE FORMATION	BLISWORTH, COSGROVE, OUNDLE, RAUNDS, PURY END, STANWICK; local	limestone, shelly, sparry, or micritic
RUTLAND FORMATION:		
Wellingborough Limestone Member	local	sparry oyster lst; sandy lst
(=Taynton Limestone Fm in SW)	HELMDON; local	shelly oyster limestone
Stamford Member	KINGSTHORPE WHITE SANDSTONE	sandstone; some carbonaceous
LINCOLNSHIRE LIMESTONE FM:		
Upper Lincolnshire Lst	[ANCASTER, BARNACK, CLIPSHAM, KETTON, STAMFORD, WANSFORD] KING'S CLIFFE, STANION, WELDON	oolitic limestone; some shelly; some sparry
Lower Lincolnshire Lst	local	sandy limestone; sparsely oolitic
	COLLYWESTON 'SLATE'	sandy limestone, fissile
GRANTHAM FM		
NORTHAMPTON SAND FORMATION:		
Duston Member:	DUSTON, EYDON, HARLESTONE, NORTHAMPTON, CHURCH STOWE	ferruginous sandstones
'Pendle'	BOUGHTON, DUSTON, PENDLE, DUSTON SLATE, KINGSTHORPE PENDLE, MEARS ASHBY, PITSFORD; local	sandy ferruginous limestones or calcareous sandstones; some crinoidal
Corby Ironstone Member:		
upper calcar. ironst.	DUSTON and HARLESTONE ('ROUGH RAG')	sandy ferruginous limestone
main oolite ironst.	BRIXWORTH, FINEDON, WELLINGBOROUGH; local	oolitic limonitic ironstone
lower calcar. ironst.	BRIXWORTH, COTTINGHAM, DESBOROUGH, GLEN HILL; local	sandy calcareous ironstone
WHITBY MUDSTONE FM	(bricks)	
MARLSTONE ROCK FORMATION	BADBY, BYFIELD, [GT TEW, HORNTON], STAVERTON; local	calcareous ironstone, ferruginous limestone

Fm = Formation; Lst = Limestone; [outside the county]

CHAPTER ONE

Stone – in and above the Ground

'Even the meaner Houses . . . for several pretty large Tracts of the County . . . have Walls
built all of Stone: a neat, secure and commodious way of Building, deny'd to some other
Counties. All the way from . . . one end of the County to the other, we meet with
Quarries; here of White Stone, there of Red: here of Freestone, there of Ragg. By which it
appears how plentifully we are stor'd with Quarrystone. Just uncovering the Earth in
some Places, they take up Stone that's every way fit for Building without almost any
Labour of the Mason; the Stone being naturally in the Earth as if it lay in a Wall.'

JOHN MORTON, *The Natural History of Northamptonshire* 1712.

The modern traveller passing through Northamptonshire sees very little of the county. The
M1 motorway descends blandly from the relative
heights of Salcey Forest (130 metres) towards the
valley of the Nene (70 metres) and then runs for the
most part up the valley of clay and gravel towards
Watford Gap, once a southerly escape-route for great
volumes of meltwater during the Ice Age. The A14
cuts a slice through the geological history of
Northamptonshire, but the grassy cutting shows
nothing to distract the motorist. It is true,
Northamptonshire's geology almost everywhere lies
hidden beneath a pastoral landscape of fields,
valleys, and gentle hills. The traveller will not be
aware of a county endowed with a wealth of stone.
There are no craggy outcrops, few roadcuts of rock,
and little visible nowadays by train. Yet this is a
county of mellow stone-built villages and market
towns, which merit a detour from the by-pass and
further exploration on foot. Though many a village
is surrounded by twentieth-century brick, its real
character lies in the stone buildings at the centre - the
church, the Hall, the manor house, local hostelry,
farmhouses and cottages.

The villages, moreover, are remarkably varied –
most striking being the colour of the stone, from the
brown-toned villages of Staverton and Badby where
the Nene rises in the western uplands, the deeper rust
of Hardingstone and Little Houghton and the rich
brown of Ecton, to the many pale grey limestone
villages overlooking the Nene valley, from Grendon

1.1 Middleton Cheney has attractive brown stone houses
around the church; The Gables (*c.*1690), with mullioned
windows, is built of ironstone from the Marlstone Rock
Formation (Lower Jurassic).

to Fotheringhay, Nassington, and Yarwell. There is a pattern to the local building; it is unequivocally linked with the underlying geology.

Geologically Northamptonshire is a north-easterly extension of the Cotswolds, lying in the swathe of Jurassic rocks that stretches across England from Dorset to Lincolnshire, narrowing towards the Humber, and widening once more in North Yorkshire. The Cotswold Hills are shaped by the thick cream limestones of the Middle Jurassic; the rock emerges visibly in the escarpments of Crickley, Leckhampton, and Cleeve Hill. The plentiful supply of good building stone has provided the picturesque architecture of the Cotswold villages with pale limestone masonry, mullioned windows, moulded door-cases, and stone-slate roofs.

Towards north Oxfordshire and south-east Warwickshire the Jurassic geology changes. Within the Lower Jurassic, an iron-rich limestone (the Marlstone Rock Formation) becomes prominent, capping the escarpment of Edge Hill, providing the warm brown building-stone of the Banbury area, and continuing into western Northamptonshire

1.2 Paine's Cottage in Oundle has a long history. Medieval in origin, with a seventeenth-century bay, in 1801 the property was left as almshouses by the bequest of John Paine. It is built of local Oundle limestone (geologically known as the Blisworth Limestone Formation), with Weldon Stone (Lincolnshire Limestone Formation) as quoins, and also the Elizabethan pinnacled feature reputedly from Kirby Hall.

(1.1). The Middle Jurassic also has distinctive geology; in place of the thick limestones of the Cotswolds are more varied rocks, with yet more iron-bearing, rusty-coloured sandstones and iron-stones, which are conspicuous in local building over a wide area of the county. There are limestones too, but they are not the massive scarp-formers of the Cotswolds – rather they occur intermittently among softer sediments, each limestone unit having its own distinctive character, and providing recognisably different building-stones (1.2).

In 1712 the Reverend John Morton published his *Natural History of Northamptonshire*, a remarkable work. His detailed observations and discussion ranged from springs and soils to flora and fauna, fossils and antiquities – and he was particularly interested in stone. He reported on the quarries, those being actively worked and those known by him to be ancient; he described the first local geological sections; and he recorded where the various types of stone had been used. Morton therefore provides a historical milestone: Northamptonshire stone as seen about 1700.

Morton himself was a remarkable man, a classical scholar who was well-read in the newer works of a world awakening to scientific observation and enquiry. Born in 1671 at Scremby, not far from Skegness in Lincolnshire, he attended Oundle School in Northamptonshire from 1686, and graduated at Emmanuel College, Cambridge, in 1692. He became first Curate, later Rector of Great Oxendon parish, and during these years he travelled extensively throughout the county ('unless Saturdays and Sundays'), observing, making notes, and collecting material for his book, which was inspired by Plot's *Natural History of Oxfordshire* (1677) and Woodward's *Natural History of the Earth*. He corresponded with many of the learned men of the day, including Sir Hans Sloane, Dr Woodward, and Mr Edward Lhwyd, Keeper of the Ashmolean at Oxford, about fossils and other natural curiosities, and specimens were exchanged. In 1703 he was elected Fellow of the Royal Society. Morton died at Oxendon in 1726, aged 55, and Sir Hans Sloane instigated funding for the inscribed stone in the chancel.

The work of historian John Bridges (a contemporary of Morton) who died in 1725, was eventually published in 1791, and certain quarries are mentioned in it. George Baker's *History* (1823-30 and 1836-41) then took the record into the early nineteenth century; but these works focus more on local history. For the purpose of this book we look to two geologists: one is Samuel Sharp, a celebrated local geologist, antiquarian and numismatist. He was born in Hampshire in 1814, but after his father's death the family moved to Stamford, where he later assisted his stepfather, owner of the Stamford Mercury. When he came to Northampton in 1857 he lived in Dallington Hall, near Duston, and his detailed accounts of Northamptonshire's geology were published by the Geological Society of London in 1870 and 1873. Several fossils were named after him – the brachiopod *Kallirhynchia sharpi* is well known to geologists in the county. He died at Great

Harrowden Hall in 1882.

The other notable local geologist was Beeby Thompson, born at Creaton in 1848, who became Headmaster of the Technical College in Northampton. His geological observations spanned many decades, his prolific writings appearing in the *Journal of the Northamptonshire Natural History Society and Field Club* from the nineteenth century until he died in 1931. Both these gentlemen saw the geology of Northamptonshire opened up in the numerous working brick-pits, in the extensive quarries dug to extract the newly discovered ironstone, and in the quarries exploited by the Victorians for building-stone. Sharp was one of the founders of Northampton's Central Museum, and Thompson's valuable, nowadays unrepeatable, fossil collection is kept there; some are on view in Abington Museum.

Much information exists tucked away among the assorted documents of local history – such as building accounts, estate records, and old maps, and will continue to be discovered by local historians. But many old stone buildings are not documented. In a village the oldest building is usually the parish church, which holds the key to stone building in the medieval period. A copy of Pevsner provides a useful appreciation of the architectural features; but a great deal can be gained by looking at the building, at its masonry, and the stone of which it is built. This book aims to show what to look for in stone, from a general view such as colour, even visible fossils, to recognising the main components of a stone with the aid of a hand-lens. It is possible, with practice, to assign a stone to the geological formation from which it came, but only a few types can be traced to known quarries. Of course, precise identification for, say, conservation projects, sometimes requires detailed examination of samples in geological laboratories, by the experts.

Northamptonshire has, in the past, been largely self-sufficient in its stone sources. Obvious exceptions are exotic facings for the shop fronts and buildings of town centres; however, they are not the subject of this book. In the villages of the county you will find few examples of stone from outside it, although close to the county boundary imported stone sometimes makes an appearance. The coming of the railway in the mid-nineteenth century enabled stone to be brought economically from greater distances, and limestone from Bath, for example,

1.3 Rubblestone: roughly dressed local Marlstone Rock (with fossils) in Daventry's old Grammar School (1600).

1.4 Freestone: as ashlar masonry in well-squared sawn blocks, with little mortar, and also for mouldings of windows and gables. Two limestones were used here at Lilford Hall (1635; see also 11.10), the gable in Lincolnshire Limestone (from Weldon, like the windows), the lower masonry in local Blisworth Limestone, which was here suitable for ashlar.

appears as lintels for quite modest brick housing. But for most older buildings we are looking at indigenous Northamptonshire stone.

First, we should consider the way stone is used in buildings. There are, in the main, two categories of building stone, irrespective of rock-type. One is common walling stone, roughly shaped, or hammer-dressed, and known as rubblestone (1.3). Walls are built up in courses if the size of blocks is sufficiently even, or otherwise (especially in old church walls) as

random rubble, set with a lot of mortar. The other is stone specially selected for particular purposes, such as well-dressed or sawn squared blocks, known as ashlar, or carved mouldings around windows and doors. These generally require freestone (1.4), ideally available in large blocks, which can be cut 'freely' in any direction. Also in the category of 'selected stone' we would include large blocks for quoins, stone slates for roofing, or slabs for paving. Specially selected stone may not necessarily be local, but will have been brought from the nearest suitable source, taking into account the high cost of transporting stone from the quarry to the site of the building. Northamptonshire had sources of several kinds of freestone, as well as stone roofing slates and paving, so even special orders could be supplied within the county. A type of masonry sometimes confused with rubblestone is rock-facing; the selected stone has in fact been carefully shaped, using a pitching tool, and laid in well-defined courses (1.5). When looking at buildings it is helpful to distinguish between the

1.5 Rock-faced masonry, popular in the nineteenth century, is well-tooled and coursed. Two types of stone were selected for this farmhouse (1872) in Mears Ashby, both from the Northampton Sand Formation: a hard brown sandstone (for rock-facing), and golden sandy limestone (a freestone, quarried locally) which was here used for quoins.

1.6 This ashlar house in King's Cliffe was built by a quarry owner in 1750, of the local King's Cliffe Stone (like Weldon Stone, it is an outcrop of the Lincolnshire Limestone).

different categories of stone. Specially selected stone begs the question, where has it come from? Rubblestone, on the other hand, usually reflects the local geology. Many buildings are combinations of the two: for example, rubblestone with freestone dressings (quoins and window surrounds) or an ashlar-faced front with side walls of local rubblestone. Of course, villages close to sources of freestone may have houses of dressed stone which is nevertheless, local (**1.6**).

Morton used the terms 'Ragg' and 'Raggstone' synonymously. He contrasted 'Ragg' with 'Freestone', the distinction being made from the viewpoint of how the stone could be worked. He noted that many varieties of building-stone would be called 'Ragg' if they were not freestones, and over much of the county 'there was scarce a Lordship without a Pit of the ordinary rougher Sort of Raggstone'. It is generally a hard stone, or otherwise unsuitable for smooth dressing, and would be used for rubblestone walling. The term 'rag' was (and is) also used for fossiliferous, sparry limestones, some of which could indeed be worked into quoins and carved features. Barnack Rag is probably the best known, a tough, shelly limestone, but able to be beautifully carved, as can be seen at Ely Cathedral. A

1.7 A monument in the church at Deene, made of Weldon 'Marble', a variety occurring within the Lincolnshire Limestone at Weldon. It is full of fossils, including a long screwlike gastropod, *Nerinea*, in a matrix of crystalline calcite.

hard bed occurring between two beds of freestone at Weldon was listed by Morton as 'Welden Ragg', and a quarry at Raunds was well-known for 'Rance Ragg'; these 'Raggs' were of such good quality as to be worth more than 'so mean an Appellation', being 'well-deserving of the name of Marbles . . . when wrought to the Smoothness they are capable of'. They are attractively fossiliferous, with shells set in a matrix of crystalline calcite (1.7 ; and see Chapter 11).

Sedimentary rock, having accumulated in the distant geological past as layer upon layer of fine particles, generally retains signs of this layering when stone is seen in the ground, in the form of beds separated by perceptible partings (bedding planes). One can appreciate that rock is rarely a homogeneous material. Deposition of the original

1.8 Sedimentary rock in layers, or beds, separated by bedding planes. These beds in the Blisworth Limestone Formation are above the level of the main building-stone once quarried underground, at Cosgrove. The site has since been largely infilled. [c.SP784420].

1.9 Beds of Blisworth Limestone in the working quarry at Pury End are traversed by a pattern of vertical cracks (joints). They are quarried by hydraulic excavator, without blasting. Blocks are then cut by hydraulic guillotine, for rubblestone building and walling [SP707460].

1.10 Cross-bedding, here seen in limestone: the shelly particles having been laid at an angle to the main bedding by currents on the sea-floor. The quarry, in the Blisworth Limestone at Lilford Lodge Farm, near Oundle, has been infilled.

sediment (say on the sea-floor) was governed by factors such as the type and grain-size of the particles, the amount of sediment, and depth and movement of water; any of these could be subject to change, even quite suddenly, so that a bed is seldom exactly the same as the one below or above (see 1.8). A good bed of workable freestone can occur among various beds that are only suitable for rubblestone. The thickness of the naturally occurring beds will determine the size of quarried blocks, and so will the spacing of perpendicular cracks (joints) (see 1.9). Beds of rock sometimes also contain fine layers at an angle (cross-bedding), formed by currents on the sea floor (1.10). Near the ground surface, rock may be affected by weathering. A succession of limestone beds, for example, can be shattered by frost within a metre or so of the surface, and sandstones reduced to sand by removal of the matrix around the grains. For this reason, some of the most satisfactory building-stone quarries are those having a protective overburden, such as clay.

Around 1700, according to Morton, there were a great many quarries, supplying stone for fine architecture and modest dwellings. Our surviving buildings need sources of similar stone if they are to be conserved, but most of the original quarries have disappeared, and only four quarries are being worked (see 2.13); the mason's work continues in Northamptonshire, using modern techniques and traditional skills (1.11, 1.12), but most stone now comes from outside the county.

This book sets out to describe the varied building stones of the county in relation to the rock that lies below the ground. It therefore begins with some geological background in Chapter 2 (and unfamiliar terms are explained in the Glossary). The succeeding

1.11 Modern stone-working machinery: Jon Aldwincle operates a computer-controlled saw at a Weldon stoneworks.

1.12 Traditional skills of the 'banker mason': an apprentice working at the stonemasons' bench at Weldon.

chapters follow in order of geological age of the rock, beginning with the stratigraphically oldest in Chapter 3; they will also be found to have a regional association. Traditional stone-working practices are mentioned in Chapters 8 and 9, but the reader is referred to *English Stone Building* by Clifton-Taylor and Ireson for a first-hand account of the work of the mason. For architectural description of buildings, Pevsner is a good companion, and there are detailed accounts in works of the Royal Commission on Historical Monuments (RCHME).

The following pages introduce new ways of looking at stone, closing in from the more familiar architectural appraisal, to the masonry, and to features visible in the stone (sedimentary structures, or fossils); then to the less familiar, often surprising view of the stone as magnified with a hand-lens. A word of caution is needed for stones that look alike. Alec Clifton-Taylor's excellent *Pattern of English Building* introduced the idea of regional distribution of materials and style, but even he mistook many of the Northampton Sand villages (including Rockingham; see 4.11) for Marlstone. (He also mistakenly thought that Helmdon Stone at Easton Neston was Marlstone.) Errors are often avoidable by using a hand-lens, but similar-looking rocks do appear at different times. Then the context should be considered, in relation to available geology, the

distance of possible transportation, and the type of building, or use, for which the stone was chosen. Another caution is needed with regard to the use of stone reclaimed from elsewhere – a common practice, as stone has always been a valuable commodity. Reclaimed sources are sometimes documented, but there can be clues in a building, such as obviously non-local material, or the random presence of burnt blocks (see Chapter 12).

This study of Northamptonshire building-stones has continued over many years, for a long time with Professor John Hudson, in connection with the classes we conducted at the Leicester University Centre in Northampton, and Knuston Hall. The local rubblestone of most villages is found to correspond with the nearest outcrop of useful stone, usually occurring within about a kilometre; hence village stone can be 'read' rather like a geological map. Chapter 2 introduces the pattern of broad outcrops of building stone across the county, and the geological reasons for it. Villages at a distance from stone sources, often on clay, are built mainly of brick. Some mud-walling is seen, along with stone, in areas of soft-weathering Northampton Sand (Ravensthorpe, Guilsborough, and Hollowell). This book is not a gazetteer, but a guide to looking at stone. A selection of examples are given, but the reader will find there is still much to discover.

CHAPTER TWO

Introducing the Geology

Northamptonshire is underlain by Jurassic rocks, which within the county are between about 195 and 150 million years old, a succession of varied sediments originally deposited beneath the sea or on coastal plains, which eventually emerged as part of the British landscape during the 63 million years of the Tertiary era; the eroded landscape later becoming covered by material deposited by ice-sheets within the last 2 million years. The glacial deposits now remain as only a partial cover, and gravels derived from them lie in the valleys, with other more recent river deposits. Building stone, iron ore, sand, and brick-clays have been obtained from the Jurassic rocks; the glacial boulder clay has also been used for bricks, and gravels for aggregate. The map of the county shown on the opposite page (2.1) is a compilation, much reduced, of the geological maps published by the British Geological Survey, mostly on a scale of 1:50,000 (they are listed with the Bibliography on page 120).

The geological map may look complicated, but the structure in Northamptonshire is really rather simple. The pattern is very much like a familiar contoured map of the landscape: rivers and streams can be picked out by the river deposits along the valleys (shown in pale yellow), the most obvious here being the River Nene and its tributaries. The relatively high ground is occupied by the remains of the cover of glacial till (boulder clay; shown pale blue), though it has gone from the western uplands. Continued erosion of the valleys has exposed the underlying Jurassic rocks which appear rather like contours as successively older layers are reached towards the bottom of a valley (there partly covered by the recent river deposits). The succession of Jurassic rocks is given in the key to the map, from the oldest at the bottom; new names have been introduced in recent years, but currently available maps mostly give the old ones (they are given in brackets here).

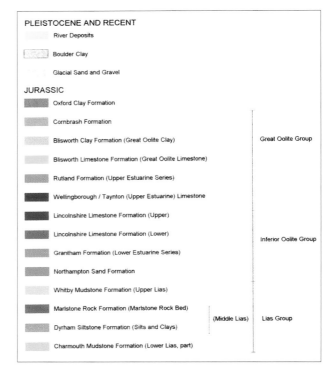

Opposite page: **2.1** An overview of the geology of the county. The course of the River Nene can be seen by the river deposits (pale yellow), from its source in the west, and its tributaries coming in mainly from the north, to where it leaves the county in the north-east. The river system of the Tove can be seen in the south. Boulder clay (blue), deposited in the Ice Age, remains on relatively high ground, but much eroded, and has gone from the western uplands. These deposits, with the associated gravels (pink), are all superficial. They overlie the Jurassic rocks which have been reached by erosion in the system of river valleys. The main towns, shown for reference, are B Brackley, C Corby, D Daventry, K Kettering, MH Market Harborough, N Northampton, S Stamford, T Towcester, W Wellingborough. (Note that in the south the compilation includes map data of different vintages, 1968 to 2002.) For local details, larger-scale maps of the British Geological Survey are recommended.

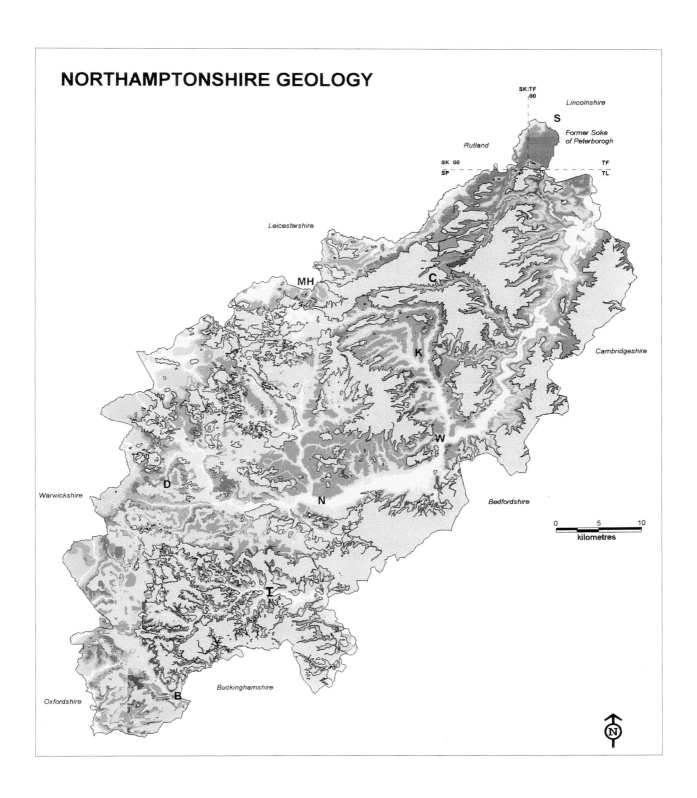

NORTHAMPTONSHIRE GEOLOGY

Across the county, there is a broad pattern in the Jurassic outcrops. The earth movements that uplifted the Jurassic rocks tilted them slightly to the south-east, bringing to the surface the oldest Lias in the west and successively younger rocks towards the east. Across the width of the county the slight tilt is largely responsible for the regional distribution of different rocks – reflected most noticeably in the differences in building stones from one side of the county to the other (see **2.13** and **2.15**). At the same time each outcrop in detail has the more intricate 'contour' pattern typical of nearly horizontal strata.

Larger-scale geological maps also show 'faults', which are lines (representing planes) along which the rocks are relatively displaced. There are not many of them, but they account for the abrupt juxtaposition of strata from different stratigraphic levels. The Kettering map (171) shows several parallel faults from just south of Corby, running east-south-east for more than 12 kilometres towards Aldwincle, within which the rocks have been displaced downwards by many metres in a structure known as the Stanion-Aldwincle Trough. Another trough-like structure (on the Towcester map, 202) occurs near Stowe-Nine-Churches, with downfaulted Blisworth Limestone at outcrop (and a source of building stone) far from its usual place. Such displacements probably have a deep-seated origin, and occurred before the deposition of the glacial material, which partially covers them. But rocks have also been disturbed by more recent superficial movements, not usually shown on geological maps, generally initiated by the creeping of clays towards the valleys, causing overlying harder beds (particularly Northampton Sand) to bend over the valley sides in the form of a saddle, opening cracks or fissures, or to be displaced by small faults, a common feature of Northamptonshire geology.

NORTHAMPTONSHIRE IN THE JURASSIC

Two hundred million years ago when the geography of the globe was very different, the present continents were joined together in a vast land-mass known as Pangaea. Prior to the Jurassic – in the Triassic period – the land-locked region of northern Europe was a hot, mainly arid red desert, where

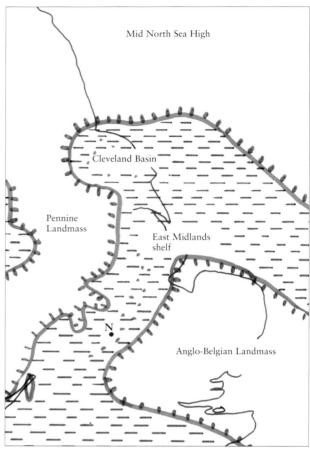

2.2 The geography of eastern and central England during the deposition of the Northampton Sand Formation, *c.* 179 million years ago (Ma). Areas of land are outlined in brown, and shallow sea shown in blue, with ironstone deposition in red. Notice the Anglo-Belgian Landmass, the western edge of it passing through Northamptonshire close to the present Nene Valley. N marks Northampton.

Adapted from Bradshaw *et al.*, with permission from The Geological Society.

deposits of sands, pebble-beds and red muds accumulated in low-lying areas (these are the red-coloured sediments we now see exposed in riverbanks and roadcuts in the English Midlands north-west of Northamptonshire). Pangaea was embayed in the south-east by an ocean known as Tethys. During the Jurassic, rifting of the crust began to divide Pangaea, opening up new seaways which gradually flooded much of what is now northern Europe.

In Britain the Jurassic sea crept over the low-lying areas of the Triassic landscape and lapped around the higher ground of the older mountains of Scotland and the north of England; south-west England and part of Wales remaining as islands. The old Caledonian mountains that stretched from Scotland to Scandinavia became divided by the sinking of the Viking trough in what is now the northern North Sea, but there remained an area, at times submerged, known as the Mid-North Sea High, which lay to the north-east of Yorkshire.

A further landmass, which is important to our area, lay across what is now East Anglia, extending into Belgium, known geologically as the Anglo-Belgian Landmass (2.2). The north-western shore of this landmass ran through Northamptonshire, more-or-less along much of the present valley of the River Nene, continuing towards Towcester and the south-west of the county. The shore of the landmass at times shrank back as the Jurassic sea crept further east, at other times the landmass expanded and the sea retreated. Such fluctuations would be caused by changes in global sea-level and more local movements in the underlying crust. As the depth of the sea varied, so the type of sediment accumulating in this area changed; lower sea-level led to shallowing and even to erosion of existing Jurassic deposits.

The temporary margin of the landmass can be recognised in the geology of the Middle Jurassic: the Northampton Sand Formation, for example, deposited in the sea across part of Northamptonshire, is missing to the south-east of Rushden, Irchester and Towcester (see 4.1). In the Jurassic the English Midlands lay some 20 degrees of latitude south of its present situation; the climate was as warm as say, the present Gulf of Mexico; and the variably shallow seas supported an abundance of life.

THE SEDIMENTS

Rivers from the uplands around the area of the British islands brought the products of erosion down to the sea, spreading out layers of sand, silt and mud. Sand is carried by swiftly moving water, and is soon deposited near the shore as the waters become sluggish; sands are mainly composed of particles derived from older rocks, the most common survivor being grains of quartz, but flakes of mica and small amounts of other minerals can be present. Mud is composed of finer particles, including newly formed clay minerals, and can be carried out to settle in quieter waters, not necessarily very deep. Calcareous deposits later to become limestones are formed as accumulations of shelly fragments (remains of the hard parts of once living organisms), or of fine calcium carbonate 'mud', precipitated in various ways from sea water.

Some limestones contain visible spherical particles formed by precipitation of calcium carbonate; they have a concentric internal structure and are known as ooliths (since they resemble eggs, as in herring roe). Rocks with these structures are described as oolitic. Ooliths are forming today, collecting in shoals, scoured by channels, in the warm waters of the Bahamas. Pure limestones develop in areas free from incoming sand or mud, but impure sandy or muddy limestones are common.

Ironstone is an unusual sedimentary rock but common in the British Jurassic. It contains a high concentration of iron-rich minerals formed on the sea-floor, either in the form of ooliths not of calcium carbonate but of an iron-aluminium silicate mineral, berthierine (once thought to be chamosite), or as fine crystals of the iron carbonate, siderite. Near the land surface today, ironstones become oxidised on weathering, and change from originally greenish-grey rocks to rusty brown, by the development of the hydrous iron oxide, limonite. The origin of these peculiar rocks is not fully understood, but they contain remains of bivalves and other creatures that inhabit a shallow sea. Ironstone formation sometimes follows a period of lower sea level, and the influx of iron perhaps came from erosion of an area of exposed lateritic Triassic red rocks.

FROM SEDIMENT TO STONE

Though sediment accumulates as particles or grains, it can become indurated by post-depositional processes. Burial under the weight of overlying sediment causes compaction, particularly in clays. But more granular, permeable material allows movement of fluid (groundwater with dissolved chemicals) through the pore-space, which eventually precipitates mineral matter between the grains, cementing them into coherent (but to some extent still porous) rock.

The cementing mineral varies. Crystalline calcite (spar) is the common cement in limestones; sandstones may also be cemented by calcite, or by silica (having the same chemistry as the quartz sand), or by iron minerals altered to brown iron oxide (limonite). By natural variation in the components, the three main rock-types of local building stones grade into each other (2.3).

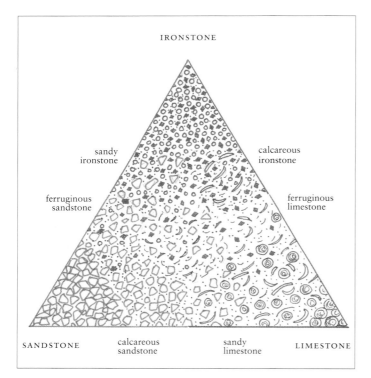

2.3 Northamptonshire's building stones comprise three main types of rock: ironstone, sandstone, and limestone; but many rocks are in fact mixtures of these constituents.
1) ironstone, in the red corner, is composed of chemically precipitated ooliths of iron-aluminium silicate (berthierine) and/or small rhombs of the iron carbonate, (siderite), with a fine-grained matrix of these minerals; all of which may be altered to limonite.
2) Sandstone, shown in green, is composed of detrital grains of mainly quartz sand (derived from erosion of older rocks).
3) Limestone, in the blue corner, may have any of several components: ooliths (with concentric structure) of calcium carbonate, formed by chemical precipitation; shell fragments (which include all kinds of fossil calcareous material); also (not shown), small pellets of lime mud, or matrix of fine lime mud; crystalline calcite (spar), the common cement in limestone and calcareous sandstone is shown here by blue dots.
Note the rock-types intermediate between ironstone, sandstone and limestone.

THE JURASSIC SUCCESSION IN NORTHAMPTONSHIRE

LOWER JURASSIC
LIAS GROUP

1. *The Blue Lias and*
Charmouth Mudstone Formations
(former Lower Lias)

The early deposits on the Jurassic sea floor were mainly grey clays but included, in the Blue Lias Formation, rhythmic layers of thin muddy limestones; this outcrop lies to the west, in Warwickshire. The outcrop of the overlying Charmouth Mudstone Formation occurs in valleys in the extreme west of Northamptonshire – near Catesby, Braunston, Kilsby, Crick and Yelvertoft. (Eastwards, these formations are hidden by overlying strata, but known from borehole records.) As much as 220 metres thick in Warwickshire, in Northamptonshire they are only 160 metres, thinning towards the Anglo-Belgian Landmass. The mudstones and limestones contain marine fossils – including bivalves and brachiopods that lived on the sea floor. Ammonites were free-swimming animals whose shells are commonly preserved as fossils; they are particularly useful to the stratigrapher because new species kept appearing, enabling one mudstone to be distinguished from another in relation to a recognised succession of ammonite zones. There also can be bones of marine reptiles (a whole *Ichthyosaurus* skeleton came from Long Itchington in Warwickshire). The limestones have been used for building in Warwickshire, but the Blue Lias has mainly been quarried to make Portland cement.

2. *Dyrham Siltstone Formation*
(former Middle Lias Silts and Clays)

These deposits, about 20 metres thick, are more sandy than the underlying Charmouth Formation, and are noticeably micaceous, buff to light grey, and in thin layers. Within the siltstones there are also somewhat coarser sandstones, as well as pebble-like concretions, and several rusty-weathering ferruginous beds, some crowded with bivalves or belemnites. The outcrop, mainly occupying the lower hillslopes of western Northamptonshire, was exposed for a time in cuttings made for the A14 road near Elkington.

3. *Marlstone Rock Formation*
(former Marlstone Rock Bed of the Middle Lias)

Above the siltstones is a hard ferruginous limestone (locally an ironstone). It may have accumulated on an off-shore shoal, with iron minerals forming among shelly detritus on the sea-floor. At the base is a pebble bed (conglomerate) derived from the erosion of earlier Liassic sediments. The thickness of the Marlstone Rock varies greatly, and is thickest (up to 7 metres) in an elongate belt from west of Banbury northwards to Edge Hill, but thinning rapidly eastwards into Northamptonshire (there is less than 3 metres at Middleton Cheney, and 1.3 metres at Milton Malsor south of Northampton). From 3 metres at Byfield, it reduces north-eastwards to 2 metres at Staverton near Daventry (**2.4**), only about one metre at Welton, and north of Crick it dies out altogether. It is missing from most of the north of the county, but about a metre was recorded by geologist J.W. Judd at Sutton Bassett, and the Marlstone Rock thickens again in east Leicestershire. Around Banbury the Marlstone Rock was worked as an iron ore, yielding on average 24 % iron metal. Burton Dassett Country Park occupies old quarries. The workable ironstone extended into western Northamptonshire and was quarried for a short time near Kings Sutton (1870s), and Astrop (1897-1924), but more successfully near Byfield (from 1917 intermittently until 1965). It has been widely used as a building stone, as described in the next chapter.

4. *Whitby Mudstone Formation*
(former Upper Lias)

Present knowledge of the Upper Lias owes much to Beeby Thompson's detailed observations in quarries and brick-pits at the end of the nineteenth century, and his meticulously recorded fossil collection in Northampton Museum (re-examined more recently by the palaeontologist Michael Howarth).

After the shallower, locally emergent interval of the Dyrham and Marlstone Rock Formations the sea extended over a wide area, deposition beginning with a thin, reddish Transition Bed and finely laminated carbon-rich 'paper shales' with an intercalated 'Abnormal Fish Bed' – a few centimetres of pale-weathering limestone containing fragments of fish (elsewhere in England there are insect remains in it too, indicating nearness to land), all amounting to no more than half a metre; overlying them are less than two metres of clays with one or more thin beds

2.4 A former quarry at Staverton in the Marlstone Rock Formation; the ferruginous limestone shows sweeping cross-bedding of shelly fragments. [SP543615].

of impure limestone crowded with ammonites. Above these beds are 50 metres of grey clays, the lower beds being generally unfossiliferous and containing pyrite nodules and selenite (transparent crystals of gypsum), the upper ones containing bivalves, gastropods, belemnites, and sometimes skeletal remains of plesiosaurs and ichthyosaurs. Three ammonite zones are known to be missing from the top, indicating a period of erosion before the Middle Jurassic.

MIDDLE JURASSIC
1. INFERIOR OOLITE GROUP

The term 'Inferior' merely indicates that this group of sediments underlies those of the 'Great Oolite Group'; in no other sense are they inferior.

1. *Northampton Sand Formation*

Classifying the various rocks of the Northampton Sand has never been easy. Sharp in 1870 recognised a lowest division (containing ironstone) overlain by 'variable beds' and 'white sands', divisions which were modified by Thompson in 1921; and by L.S. Richardson in 1925 who separated the White Sands as 'Lower Estuarine White Sands', leaving the Northampton Sand in two divisions. A complicated scheme was later devised by Geological Survey geologists Hollingworth and Taylor, but the Survey's suggested (unpublished) two-fold division, the Corby Ironstone Member and the Duston Member is

2.5 Samuel Sharp (in top hat) in Duston ironstone quarry (Northampton Sand Formation) in 1868. This was locally the thickest workable ironstone in the county, 8 or 9 metres, and was worked between 1859 and 1909; once a Romano-British settlement, now a commercial site [SP725605]. (From Beeby Thompson, 1928).

followed here; whilst noting the local prominence of 'Pendle' limestone.

Several features can be noticed about the distribution of the Northampton Sand (see the map 4.1). First, it dies out to the east and south-east, just reaching one side or the other of the Nene valley north of Rushden, and it is missing south-east of a line from Rushden towards Towcester. At some stage this was near the edge of the London-Belgian Landmass. If the Northampton Sand was deposited further east, it was subsequently removed by erosion: along this south-eastern margin, later deposits of the Middle Jurassic rest directly on the Whitby Mudstone Formation. There was also apparently some kind of structure from about Grendon to Great Doddington, extending towards Hardwick, Orlingbury and Pytchley, where Northampton Sand is also thin or missing; it has been called the Doddington-Pytchley Axis, and is visualised as a structural rise that failed to collect or retain sediment. To the east of it, the Northampton Sand

Formation comprises only the Corby Ironstone Member, any overlying Duston Member, if present, having been removed by erosion in the Jurassic. To the west of it, the Corby Ironstone is covered by quite thick variable deposits of the Duston Member. The Northampton Sand Formation becomes increasingly sandy to the west, where ferruginous sandstones predominate, but there is thick oolitic ironstone locally at 'old' Duston (see below), and ironstone underlies the sandstones in places, such as Culworth.

A. Corby Ironstone Member

The first deposits, resting unconformably on the eroded Whitby Mudstone, were sands rich in iron minerals, enclosing at the base a collection of derived pebbles, and black phosphatic nodules that were pitted with borings by the mollusc, *Lithodomus*. The lowest sandstones were often followed by sandy calcareous ironstones with siderite; neither type was workable as iron-ore, and both were called 'bastard stone', lying below richer oolitic ironstone. Similar calcareous rock also formed in the upper part of the Ironstone Member at New Duston.

The ironstone formed in an elongate zone (up to 25 kilometres wide about Kettering), stretching north-eastwards from Towcester to Corby, and then north to Grantham; The thickness of the deposit varies greatly, the maximum occurring at 'old' Duston, west of Northampton, where Sharp measured 9 metres (2.5); but he found less than 3 metres in a stone-pit in Northampton less than 4 kilometres away.

The ironstone is commonly oolitic, with ooliths of iron-aluminium silicate (berthierine) in a matrix containing the carbonate, siderite; some is a siderite mudstone, composed almost entirely of fine-grained siderite. Some layers are shelly, with bivalves often as hollow moulds. These rocks were often greenish grey when seen fresh in the deeper working quarries, but at outcrop they are always weathered rusty brown, due to the formation of hydrated iron oxide (limonite), often rather soft and ochreous, but also as harder veins of the dark brown crystalline form, goethite. The pattern of weathering is one of the interesting features to see in ironstone outcrops (in Twywell Local Nature Reserve, for example). It proceeds from the rectangular joints by chemical migration, producing layers of tough goethite in the form of 'boxstones', which can be 15 to 25 centimetres across, often several inside one another, with softer leached material inside (2.6). (In the nineteenth century both Sharp and Thompson referred to ironstone with boxstone structure as 'cellular', and Baker mentioned 'that peculiar conformation . . . called an Eagle's Nest'.)

The iron ore was worked by the Romans, and medieval charcoal-smelting took place in Rockingham Forest, but it stopped when wood was needed for building Tudor ships; surprisingly, the existence of iron ore was then mostly forgotten for three hundred years. Even John Morton in 1712 did not recognise it, and on finding slag he suggested that ore had in the past been brought from elsewhere to smelt with local charcoal. The Victorians rediscovered the ironstone; it was exhibited in 1851, and new railway cuttings revealed its wide extent, leading to the growth of Northamptonshire's iron industry which from 1853 was to last more than a hundred years. By 1980 it had declined, the last of the Corby quarries closed, and much of the landscape was restored. Throughout its outcrop the Corby Ironstone Member provided several kinds of building stone, which are described in Chapter 4.

2.6 Ironstone of the Northampton Sand Formation, exposed at Raunds [TL009731]. The iron minerals are oxidised by weathering to brown limonite, and concentric box-stone structures are formed between joints.

B. Duston Member (former Variable Beds)

The Corby Ironstone in central Northamptonshire is overlain by ferruginous sandstones (locally also including sandy limestone) which are as much as 8 metres thick between Duston and Harlestone, and have been extensively quarried for building stone (2.7 – 2.9) (Chapters 5 and 6). Part of the quarry in New Duston described by Samuel Sharp in 1870 still exists, although much of it has been infilled and is occupied by a housing estate; it is a Regionally Important Geological Site (RIGS) [SP713627] (2.9). The section can be summarised as follows (Quarryman's terms in brackets, for Sharp's details see 5.9):

	Metres
Thin boulder clay and soil	
Rutland Formation:	
7. Pale grey sands with rootlet marks near base	1.50?
Northampton Sand Formation:	
Duston Member:	
6. Brown sandstone with rootlets	0.25
5. Yellow to brown sands and limonitic sandstones, cross-bedded ('Ryelands')	2.25
4. Light buff calcareous sandstone and oolitic sandy shelly limestone, strongly cross-bedded; some fissile ('Pendle')	1.20
3. Orange-brown limonitic sandstone ('Yellow Building Stone')	1.64

Below present ground level (as seen 1979-1985):	
Loose scree obscuring lower part of 'Yellow Building Stone'	0.40
2. Limonitic sandstone as broad level ('Hard Brown')	0.25
Rubble slope, not exposed	0.50
Orange and limonite-seamed sandstone partly exposed	0.70
Corby Ironstone Member:	
1 .Ferruginous, calcareous sandstone, green-hearted, with limonite seams, bivalves (*Astarte*), and belemnites ('Rough Rag')	1.20 seen

The two divisions of the Northampton Sand Formation (not yet formally defined) are given here as divided by Richardson. The sandstones are composed of generally angular grains of quartz, with some feldspar, in a matrix of limonite, some with calcite. Within the sandstones the calcareous rock known locally as 'Pendle' is relatively thin at Duston, and in the present Harlestone quarry, but thickens to the east (it is about 5 metres at Boughton and Pitsford), where it takes the place of the lower sandstones seen at Duston, and overlies the Corby Ironstone. This limestone lagoon, inhabited by bivalves, crinoids and corals, reached Mears Ashby;

2.7 The Harlestone quarry opened by Eli Craddock on the Althorp Estate in 1892 (he is perhaps in the centre, with his daughter). The quarry, in Northampton Sand sandstones, is still being worked further west (see 2.8), this part having now been infilled. (Northampton Museums).

2.8 Cross-bedded brown sandstones of the Northampton Sand in the working part of Harlestone quarry [SP708638].

the area is roughly outlined (III) on the map (see 4.1), and its calcareous building stones are described in Chapter 6. The 'Pendle' is missing from central Northampton, and from Northampton Sand outcrops further south and south-west, until calcareous rocks are seen again near Newbottle [SP517365]. Brown, ferruginous sandstones (2.8)

2.9 The former building-stone quarry at New Duston is a Regionally Important Geological Site (RIGS), where most of the proposed 'Duston Member' of the Northampton Sand Formation can be seen [SP713627]. The thick lower and upper sandstones are separated by 1.2 metres of sandy limestones ('Pendle'). In the top of the face is about a metre of pale sand of the Rutland Formation (and the downwash tends to coat the otherwise brown upper sandstone).

overlie the 'Pendle' limestone. Bedding surfaces are often ripple-marked, and pieces of drifted wood sometimes found, all indicating sedimentation in shallow water close to a shore.

In the east and north of the county the variable Duston Member is thin or missing altogether (see above). At Finedon, less than a metre of ferruginous sandstone rests on the Ironstone, while at Irchester, the Ironstone is directly overlain by later sands of probable Rutland Formation. If the Duston Member was once present here, it was removed by erosion sometime during the Middle Jurassic.

2. Grantham Formation
(former 'Lower Estuarine Series')

Overlying the Northampton Sand Formation are sands, silts and clays which often contain signs of vegetation – such as the dark carbonaceous remains of rootlets of horsetail-like marsh plants. Over much of the East Midlands there was now a marshy delta (despite the earlier name it was not an estuary), and the sea lay to the south in the southern Cotswolds (2.10). Due to the difficulty in distinguishing the Grantham Formation from the later somewhat similar Rutland Formation ('Upper Estuarine Series') the most reliable record is to be found north of Kettering, towards Corby, and north Northamptonshire, where the Lincolnshire Limestone clearly separates the two. Here the varying successive deposits of fine sand and mud are 6 or 7 metres thick. Over even short distances, the deposits vary, as the sinking swamp accumulated sediment by means of anastomosing rivulets. They are either pale grey or yellow-brown, and often contain a layer of black carbonaceous shaly clay. South-east of the area where Lincolnshire Limestone occurs (i.e. the line Maidwell – Weekley – Tansor), the extent of the Grantham Formation is uncertain. Recent studies by Fenton and others of the sands overlying the Northampton Sand at New Duston have identified microscopic algal cysts as probably belonging to the much later Rutland Formation (see below).

2.10 When sea-level dropped back, much of eastern and central England was occupied by swampy plains with marsh plants, receiving sand and mud brought by rivers from the higher ground - the deposits of the Grantham Formation (177 Ma). Land is shown brown, and shallow sea, blue. N marks Northampton.

Adapted from Bradshaw *et al.*, with permission from The Geological Society.

3. *Lincolnshire Limestone Formation*

By the time known as the Bajocian Stage (174 million years ago), rising sea level flooded part of the marshy delta, probably coming around the north of the Anglo-Belgian Landmass, and the Lincolnshire Limestone was deposited in warm, shallow water over the East Midlands Shelf. The limestone is thickest in Lincolnshire (about 40 metres around Grantham) and tapers, like a lens, in a southerly direction towards Stamford and northern Northamptonshire. It dies out before Kettering and just reaches Maidwell. (It may well have extended further, but was soon removed by the major erosive interval that followed it.) Two episodes of limestone deposition are recognised in Northamptonshire, both of them providing valuable building stones which are described in Chapters 8 and 9.

A. Lower Lincolnshire Limestone

The lower division is the more widespread, and is up to 12 metres thick at the Rockingham escarpment overlooking the Welland, thinning away south-eastwards. The first deposits locally comprise less than a metre of sandy limestones and calcite-cemented sandstones; more often, they are loose sands enclosing large, partly cemented concretionary masses called 'potlids'. They overlie the sands of the Grantham Formation sometimes without any obvious break. The locally developed sandy limestone, cross-bedded and potentially fissile, is the source of the well-known Collyweston 'slates' (see Chapter 8). It contains marine bivalves such as *Gervillella acuta*, but in beds just above, records of drifted ferns and other plants indicate land not far away. The rest of the Lower Lincolnshire Limestone is generally fine-grained and sandy, but some has a scattering of ooliths. Within the limestones are beds containing a great many *Nerinea*, which are elongate, screw-like gastropods, several centimetres long. Sandy limestones higher than the basal Collyweston horizon, with layers of orange sand, can still be seen in old pits between Collyweston and Easton-on-the-Hill. Numerous former stone-pits – around Geddington, Pipewell, Harringworth, Wakerley, Collyweston and Easton – were described by the geologists Samuel Sharp and J.W. Judd in the nineteenth-century. Around Kettering and Corby huge twentieth-century ironstone quarries exposed the geology, with Lincolnshire Limestone in the overburden, which was recorded by Taylor in 1963, but only part of one large face remains near Corby [SP928904], and another, overgrown, near Nassington [TL045970].

B. Upper Lincolnshire Limestone

The later limestones do not form an even, upper layer over the Lower Lincolnshire Limestone in Northamptonshire, but are mostly restricted to former submarine channels which scoured into the lower limestone, locally cutting through it, reaching the underlying Grantham Formation or even the Northampton Sand. The deposits in the channels are coarse-grained and conspicuously oolitic limestones, locally also containing abundant pieces of shell in cross-bedded layers, or crowded with small fossil debris. The main channelling in northern Northamptonshire occurred in a narrow belt, five to seven kilometres wide and aligned south-west to north-east, from Pipewell south-west of Corby, towards King's Cliffe and Apethope and probably extending north-east to Barnack. Channels used to be well exposed in some of the ironstone quarries near Corby (2.11), the eroded surface of the lower limestone penetrated by the borings of worms and molluscs. The Upper Lincolnshire Limestone near Weldon and Corby was up to 12 metres thick, much of it being removed in quarrying for ironstone.

After the deposition of the Lincolnshire Limestone Formation the sea retreated once again from the East Midlands and the whole area was exposed to erosion for 4 or 5 million years, reducing the upper limestones in Northamptonshire to channel remnants, and removing some of the earlier Jurassic deposits across the county.

II. GREAT OOLITE GROUP

1. *Rutland Formation*
(former 'Upper Estuarine Series')

Above the Lincolnshire Limestone is another group of sands and clays rather similar to those of the earlier Grantham Formation ('Lower Estuarine Series'), which J.W. Judd called the 'Upper Estuarine Series' in 1867; the replacement name proposed by Martin Bradshaw removes the misconception that there was an estuary. Beeby Thompson's detailed sections measured over many decades (recorded in a classic paper published by the Geological Society of

2.11 Upper Lincolnshire Limestone fills a channel that has cut down through Lower Lincolnshire Limestone and the Grantham Formation, reaching the Corby Ironstone. This used to be seen in Cowthick Quarry, Weldon, near Corby, which is now a landfill site.

London in 1930, shortly before he died in 1931) are invaluable, though his ideas have been superseded.

A. Stamford Member

Following the prolonged erosion of the Inferior Oolite rocks across the Midlands, deposition resumed in the early Bathonian Stage (about 166 million years ago) with the Rutland Formation overlying the eroded surface (unconformity). The Midlands was at first still a long way from the sea which lay to the south; local freshwater lakes, collecting black clays, were followed by more extensive pale sands of a freshwater delta with rotting plant debris and the vertical rootlets of horse-tails. The deposits known as the Stamford Member are 3 or 4 metres thick.

South of the occurrence of Lincolnshire Limestone it has always been difficult to distinguish the 'Lower' and 'Upper Estuarine' sands and clays, though they are shown separately on geological maps. Bradshaw has suggested that the soft Grantham Formation was largely removed by erosion south of the Lincolnshire Limestone's protective cover, and most of the pale sands mapped as 'Lower Estuarine' are probably the Stamford Member of the Rutland Formation. The suggestion is confirmed by identification of algal material found in beds formerly mapped as 'Lower Estuarine' overlying the Northampton Sand at New Duston, now generally recognised as Rutland Formation. Along the east side of the Nene valley the Rutland Formation oversteps the eroded edge of the Northampton Sand and rests on the Whitby Mudstone Formation. Pale sands with carbonaceous slivers and rootlet markings were indurated into grey or white sandstone at Kingsthorpe, and became the source of an interesting local building stone (see Chapter 7).

The rest of the Rutland Formation (about 6 metres) was formed by several repeated ('rhythmic') incursions of the sea, depositing sand, silt, or sandy limestone, often containing brackish-water or marine bivalves, including oysters. Each batch of marine sediment passes upwards into carbonaceous mud in which vertical rootlets indicate another advance of the shoreline. In contrast with the submerged limestone lagoon of the Cotswold area, the Midlands remained relatively high, only

intermittently inundated by the sea. The most prominent bed in the Rutland Formation is a limestone, described next.

B. Wellingborough Limestone Member (former Upper Estuarine Limestone)

A persistent oyster-bearing limestone within one of the Rutland Formation 'rhythms' becomes a mappable unit from around Geddington southwards. The name supersedes Thompson's 'Upper Estuarine Limestone' (1909) for much of the county, but south-west of Wappenham it is known as the Taynton Limestone Formation.

Near Geddington, Cranford and Burton Latimer it includes marl full of grey oyster shells (it used to be called 'pen-earth', as the oysters resembled pennies). In the former ironstone quarry that is now Irchester Country Park it is seen as a distinct metre-thick rib within the clays of the overburden (**2.12**): here it includes thinly bedded sandy limestone with bivalves (especially the small round oyster, *Placupopsis*) and urchin spines, and in places there is a hummocky layering produced by filamentous algae which trapped fine lime mud in shallow water. By Tiffield it forms a massive bed of very shelly, oyster-rich spar-cemented limestone overlying ferruginous calcareous sandstone with whole oysters, and marls with many small echinoids.

2.12 Irchester Country Park occupies an old ironstone quarry and is a Regionally Important Geological Site (RIGS) [SP914657]. The ironstone at the bottom is obscured here by landslip. Within the soft sands and clays of the Rutland Formation overlying the ironstone is the prominent ledge of the Wellingborough Limestone; cream, marly limestone of the Blisworth Limestone Formation comes in at the top of the face. (This was 1976; there is more vegetation here today.)

c. Taynton Limestone Formation (former Upper Estuarine Limestone in south-west Northamptonshire)

The same limestone becomes thicker (over 3 metres) towards the south-west, and around Helmdon it was once a well-known building stone (see Chapter 10). It passes into the more famous Taynton Limestone of Oxfordshire, though the Helmdon and Taynton limestones are not at all similar, typical Taynton Stone being oolitic with a strong spar cement, while Helmdon limestone is more granular and shelly. At Helmdon the limestone is exposed along a railway cutting and a little of it by the sawmill [SP586443]; the upper beds are flaggy and sandy with soft fossiliferous marls, but extracted blocks of the more massive limestone (no longer exposed) are mostly composed of broken shell and granular fossil debris (including echinoid spines), with laminar streaks or thin beds of finer micrite. Oysters are abundant on some bed surfaces.

The limestone is, as elsewhere, overlain by further generally non-marine clays of the Rutland Formation, but the appearance of the small brachiopod *Kallirhynchia sharpi* in the clay heralds the next marine incursion.

2. *Blisworth Limestone Formation* (*Great Oolite Limestone*)

The East Midlands was eventually inundated by the widespread advance of a shallow sea, and relatively thick fossiliferous limestones accumulated. Thompson in 1891 recorded 10 metres of the limestone at Stowe-Nine-Churches, but at Blisworth to the east the maximum is 7 metres, and in the north, at Nassington, less than 5 metres. Thompson published a very useful survey of limestone resources in 1927, and several sections, including Blisworth [SP715553] and Cranford [SP929768] were more recently described in detail by Hugh Torrens. At Irchester Country Park, the lowest limestone deposits are crowded with *Kallirhynchia sharpi*, and interbedded with soft marls, before more massive limestone beds of shell debris come in. Bivalves are common, particularly oysters – but also many others, with brachiopods, echinoderms, corals, crustaceans, and occasional plates of shark teeth. Blisworth Limestone in Northamptonshire is not often oolitic, but contains oval grains having a fine lime carbonate coating around pieces of shell (they are called superficial ooliths). In most places the limestone is partly fine-grained (micritic) and soft-weathering, but cross-bedding is common in many sections along the outcrop (see **1.10**)., and towards Oundle the rock is also shelly and sparry.

The limestone outcrop runs the length of Northamptonshire, partly covered by boulder clay, and it provided local building stone throughout this area, as described in Chapter 11.

3. *Blisworth Clay Formation*

The sea had once more withdrawn to the south, and rivers brought fine mud to be deposited across the Midlands, in many places on top of Blisworth Limestone that enclosed plants in growth position. Otherwise, plants are not usually seen in the Blisworth Clay. A thin limestone packed with oysters, formed when brackish water came in, used to be visible in the Blisworth Clay at Pury End Quarry. The clay is between 3 and 4 metres thick, but is seldom exposed, the outcrop forming a slope between the Blisworth Limestone and the Cornbrash.

4. *Cornbrash Formation*

The outcrop, mostly alongside the Nene valley, is often seen as rather rusty soil with many lumps of limestone and quite often, fossils. It was formed during the return to a shallow carbonate sea and was evidently a particular habitat for bivalves. The Lower Cornbrash, only a metre or so thick, was eroded before the Upper Cornbrash (also only about a metre thick) was deposited, taking in at the base a layer of worn limestone pebbles (some encrusted with serpulid worm tubes), and bivalves, derived from the underlying Lower Cornbrash. Where it is not rubbly, the Cornbrash can be a tough limestone. It is not much used for building, though it might be found in local walls.

KELLAWAYS AND OXFORD CLAY FORMATIONS

The lowest part known as the Kellaways Clay is less than 3 metres thick, and is overlain by a slightly greater thickness of brownish Kellaways Sand, before the incoming of the thick mud-rock of the Oxford Clay. The outcrops lie either side of the Nene valley, but are largely covered by a blanket of yet more clay – the boulder clay – so that it can be hard to tell where one ends and the other begins in the featureless landscape that continues east towards the Fens. In the past there were local brick-pits in these rocks. The last one in Raunds was worked by Tom Smith in the Kellaways beds. Major brickworks utilising the Oxford Clay have for many years been centred on Peterborough, and huge brickpits here give the geologist the opportunity to search the clay for fossils. The dark grey mud accumulated in the open sea – it was not particularly deep, and bivalves lived in the mud on the sea-floor, but it was the home of the free-swimming ammonites and belemnites, fish and large reptiles; remains of several kinds of plesiosaur and marine crocodile have been found near Peterborough and can be seen in the Museum. Younger Upper Jurassic rocks occur outside the county.

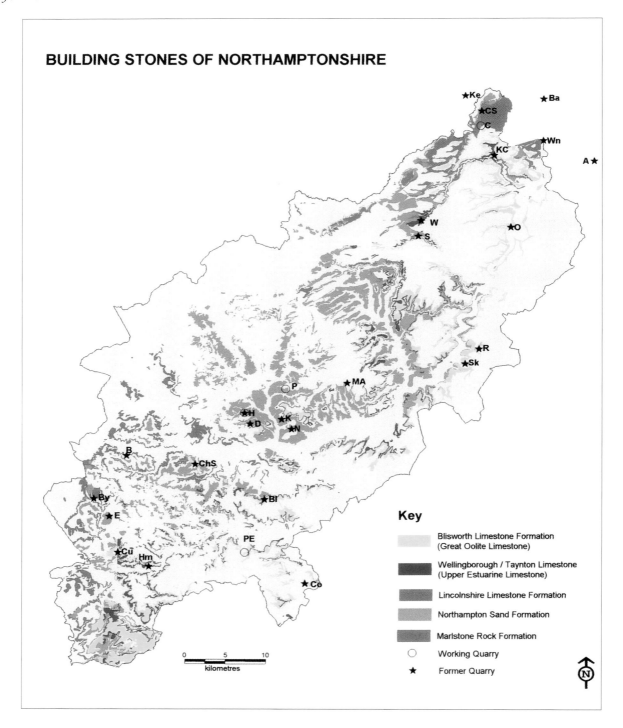

BUILDING STONES OF NORTHAMPTONSHIRE

Key

Blisworth Limestone Formation
(Great Oolite Limestone)

Wellingborough / Taynton Limestone
(Upper Estuarine Limestone)

Lincolnshire Limestone Formation

Northampton Sand Formation

Marlstone Rock Formation

○ Working Quarry

★ Former Quarry

2.13 A map of the county showing the outcrops of the chief rocks providing building stone; other rocks such as clays and sands have been omitted (this leaves out the Kingsthorpe Sandstone). The quarries indicated are named as follows: A Alwalton, B Badby, Ba Barnack, Bl Blisworth, By Byfield, C Collyweston Quarry, ChS Church Stowe, Co Cosgrove, CS Collyweston Slates, Cu Culworth, D New Duston, E Eydon, H Harlestone, Hm Helmdon, K Kingsthorpe, KC King's Cliffe, Ke Ketton, MA Mears Ashby, O Oundle, P Pitsford, PE Pury End, N Northampton, R Raunds, S Stanion, Sk Stanwick, W Weldon, Wn Wansford.

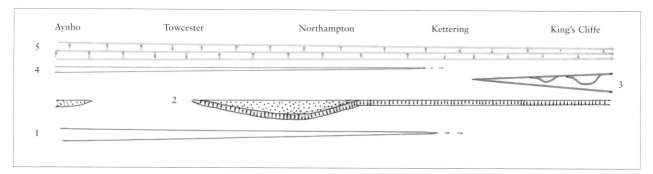

Above: **2.14:** Diagrammatic vertical slice through the geology, much exaggerated, along the length of the county, from Aynho (SW) to King's Cliffe (NE), showing how the different rocks occur. 1) The Marlstone Rock Formation is thickest in the south, and dies out in the north of the county; 2) the Northampton Sand Formation is thickest near Northampton, where the Corby Ironstone is overlain by the Duston ferruginous sandstones, while in the north-east the sandstones are missing; 3) the Lincolnshire Limestone Formation only occurs north of Kettering, thickening northwards into Lincolnshire; 4) The Wellingborough Limestone thickens south-westwards, continuing as the Taynton Limestone near Helmdon; 5) the Blisworth Limestone Formation occurs the length of the county, thicker in the south-west.

Below: **2.15** Sketch section through the geology across the width of the county, showing 1) the effect of the easterly tilt, bringing the outcrop of the older Marlstone Rock (red) up in the west, and successively younger Northampton Sand (brown) and Blisworth Limestone (yellow) to the east; and 2) the effect of erosion of the landscape, the upper rocks being removed from the valleys. (The scale is exaggerated.)

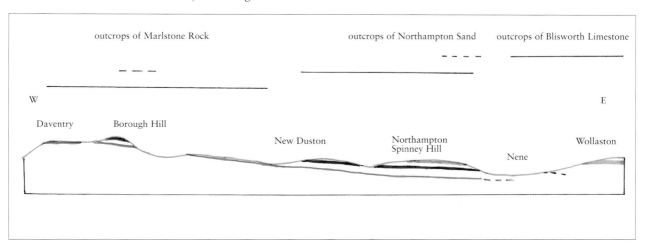

BUILDING STONES

The map on the previous page (**2.13**) shows the outcrops of the rocks used as building stones, omitting the clays and other soft rocks. The quarries shown were historic sources of freestone, for ashlar or dressed stone, and most were known to John Morton in 1712.

The distribution of the different rocks is largely dependent upon four factors –
1) the location of original deposits – which is shown diagrammatically in a section along the length of the county from south-west to north-east (**2.14**);
2) the effect of subsequent gentle tilting to the south-east – which is illustrated by a section acoss the width of the county (**2.15**);
3) removal of rocks by erosion of the landscape, affecting the pattern of outcrops (as shown in the last figure); and
4) the partial covering by glacial and recent 'drift' (see **2.1**).

It is worth remembering that rocks are accessible not only at outcrop, but by excavation through a certain amount of overburden. The main building stones (at some time available as freestone for ashlar or mouldings, or other special purposes such as roofing) are summarised in the Table on page 7.

CHAPTER THREE

The Marlstone Rock Formation in the West

The scenic uplands in the west of the county are carved in sediments of the Lias Group, capped in many places by outliers of Northampton Sand – including Red Hill and Eydon Hill to the south of Byfield, Arbury Hill (225 metres) near Badby, and Borough Hill by Daventry. Within the Lias the Marlstone Rock, about 3 metres thick, is prominent in the steep escarpment and the tributary valleys of the Cherwell in the west, from Kings Sutton to Byfield. Further north, the tributaries of the easterly-flowing Nene have cut through the Marlstone Rock around the valleys from south of Badby to Watford (3.2). In western Northamptonshire, brown Marlstone Rock is the local building stone of more

3.1 Cottages in the delightful village of Kings Sutton, built of local calcareous ironstone from the Marlstone Rock Formation.

than 30 delightful villages, lying on or close to the outcrop (3.1).

The outcrop can be traced east along the Nene valley to Harpole and to the incoming valleys around Rothersthorpe and Milton Malsor near Northampton, before the regional geological dip takes it down below the valley floor. However, this far east the Marlstone Rock Formation has become a thin layer (less than a metre), and its use as a building stone limited, while other building stone – including sandstone of the Northampton Sand – is available close by. North of Crick, the Marlstone Rock more or less dies out, and is missing from the north of the county except near Sutton Bassett. Beyond the county to the south-west, the Marlstone Rock has been an important building stone, which is still quarried between Banbury and Edge Hill.

Villages are mostly built of rubblestone, and their

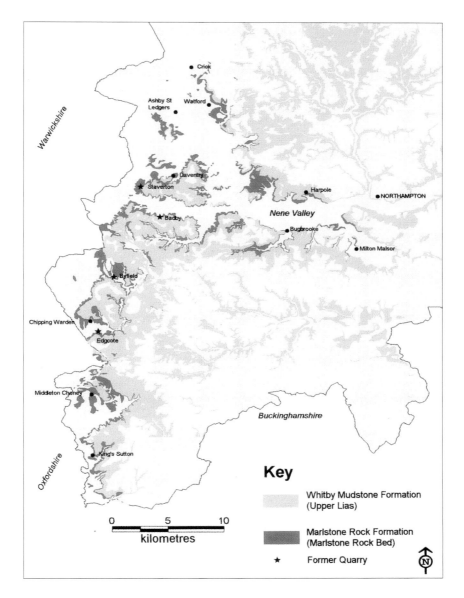

Key

Whitby Mudstone Formation (Upper Lias)

Marlstone Rock Formation (Marlstone Rock Bed)

★ Former Quarry

cottages (**3.1**) are not only aesthetically pleasing, they are often the key to the local geology, for their stone is likely to have come from the nearest stone-pit. There were once many stone-pits in the Marlstone Rock. Thompson recorded those that were known in 1888, including two south of Thenford ('though I was assured by an inhabitant that there were no stone pits about'), some near Middleton Cheney, and towards 'Chalcomb' [Chacombe] one from which the Wesleyan Chapel there had been built; he described others at Aston-le-Wall, Hellidon ('The Rock-bed is found here in large blocks . . . and would probably make a good building stone'), Welton, Bugbrooke, and at

3.2 Map of south-west Northamptonshire, showing the outcrop of the Marlstone Rock Formation (red), with the Whitby Mudstone Formation which overlies it. The following villages, on or close to the outcrop, are built of Marlstone Rock (those in brackets also have a significant amount of other material, such as Northampton Sand sandstone):

Aston-le Walls, (Ashby St Ledgers), Badby, Boddingtons, (Bugbrooke), Byfield, Chacombe, Charwelton, Chipping Warden, (Crick), (Daventry), Dodford, (Everdon), Fawsley, (Flore), Gt Purston, (Harpole), Hellidon, (Kilsby), Kings Sutton, Marston St Lawrence, Middleton Cheney, (Milton Malsor), Newnham, Norton, Overthorpe, (Rothersthorpe), Staverton, (Sutton Bassett), Thenford, Watford, Welton, West Farndon, (Woodford Halse).

Rothersthorpe – where, however, 'every quarry has been grassed over'. At Badby, Bridges had noted in the eighteenth century, 'there are several quarries of a fine blue rag stone, very hard and durable, from whence large quantities are carried into Warwickshire, for building and pavement'. The quarries were still worked a century later but, according to Baker, not so extensively. No Marlstone Rock is worked now in the county.

The stone is interesting and well worth a closer look – in barn walls, in church masonry, or buildings by the roadside. Its various shades of brown are mainly due to the natural weathering of originally more blue-grey rock, patches of which sometimes remain. It is a composite rock of iron minerals and limestone (**3.3**), and although classed as a limestone in building-stone records, in some areas it was sufficiently iron-rich to have been quarried as ironstone, especially in the south-west, around the villages of Byfield and Charwelton. Ironstone magnified with a hand-lens is often found to be oolitic.

The stone in buildings can provide graphic reminders of the Jurassic sea (**3.4**, **3.5**): in places there are sweeping lines of paler limy particles laid by currents on the sea-bed, as well as light and dark patches produced by burrowing creatures that churned up the mud (in some rocks there are visible little infilled burrows), shells of bivalve molluscs and tiny ossicles of crinoids; particularly common in Marlstone Rock walls are brachiopods and belemnites. The brachiopods are of two types, ribbed rhynchonellids and smooth terebratulids; they are roughly a centimetre or so across in any direction, their two shelly valves of calcite having remained joined together (unlike many bivalve mollusc shells); when cut across in walling-stone they may look hollow or contain crystals of calcite. They lived on the Liassic sea-floor, but the belemnites were free-swimming squid-like animals – each with an internal hard skeleton, cylindrical but tapered, bullet-like, to a point at one end; in a wall there can be elongate or round cross-sections, composed of prismatic crystalline calcite. These common fossils help to distinguish Marlstone Rock from some types of the Northampton Sand, which can look very similar, when both are weathered brown and veined by dark brown limonite.

Blocks of stone in a wall are often surprisingly varied – even when they may have come from the

Top: **3.3** The calcareous ironstone from the Marlstone Rock, as magnified with a hand-lens. It consists of green iron-minerals, partly altered to brown limonite (some ooliths are just visible); the light-coloured mineral is calcite. (Leicester University accession no.: LEIUG119480).

Above left: **3.4** Varied walling of the Marlstone Rock in Byfield, showing sediment swept by currents on the Jurassic sea-floor, and fossil belemnites (top left).

Above right: **3.5** Fossil brachiopods and driftwood seen close-up in a cottage wall of Marlstone Rock, Church Lane, Middleton Cheney.

same stone-pit – with differences in colouring, perhaps fossils in one or two blocks; they make an assemblage, from different beds or even different parts of the same bed. From one village to another the Marlstone Rock assemblages also may look rather different. In the south-west the building-stone includes ironstone, dark bluish brown at Byfield, some weathered reddish brown at Overthorpe. Elsewhere, there is stone with plum-coloured cores, tawny-weathering, which may have more siderite, and paler variants (described below). It is convenient to call all of them calcareous ironstone.

Marlstone Rock rubblestone is recognisable in all

the villages close to the outcrop, but there is also good-quality, better-dressed stone from this Formation, which may or may not be strictly local; it should be compared with the local material, and with stone from known building-stone sources. There are fine seventeenth-century houses, for example, in Kings Sutton and Middleton Cheney, with stone mullions and mouldings (see 1.1), and an eighteenth-century manor house of ashlar in Byfield (3.6). Was the stone local, or brought from further afield? Kings Sutton, Middleton Cheney and Byfield have particularly fine churches with elegant spires. Some architectural and decorative features are of limestone but the masonry of all three churches is well-dressed calcareous ironstone (3.7), as are most of the piers and arcades, and even some more delicate carving. But where were the fourteenth-century quarries? Near Edge Hill, by the Warwickshire-Oxfordshire border, are quarries that have worked the Marlstone Rock for centuries. Some are still working in Oxfordshire, producing Hornton Stone (also now Great Tew Stone) which are used for renovating brown stone buildings (Marlstone Rock or Northampton Sand) in Northamptonshire (see 3.14). But in the fourteenth century would stone have been transported from Edge Hill across to the Cherwell valley and up the steep escarpment to Middleton Cheney, or a distance of some 20 kilometres to Byfield, if good stone was available from local sources? Stone seen at Byfield is darker than

3.6 The Manor House, Byfield, an ashlar house of dark calcareous ironstone from the Marlstone Rock. (Built early eighteenth century, probably for William Hitchcock, it was sold in 1758 to Richard Chauncey of Edgcote.)

most Hornton Stone, and looks similar in cottages, the church, and the manor house, so it was probably local.

There is evidence for quarries at Byfield in the eighteenth century, though some have been destroyed by twentieth-century quarrying for iron-ore. Byfield's Enclosure Award map for 1779 shows three stone-

3.7 The fourteenth-century south porch of Byfield Church, built of dark Marlstone Rock, with limestone from Helmdon.

pits: one was south of the bridleway to Aston, one north of Iron Cross, and one on the east side of the Charwelton road; some could have been just for road repairs. But John Morton in 1712 described stone 'from Byfield Quarry, of a dark Colour, almost Black' which was used in conjunction with white limestone from Culworth, 'wrought to a considerable Degree of Smoothness, so as to nearly approach a Polish. The Halls of most of the Gentlemens Houses in that Part of the County are paved with these two alternately set in Squares, in the usual manner of paving with Black and White Marble'. An example can be seen in the hall of Canons Ashby House, which was modified by Edward Dryden in about 1710 (3.8). There are more dark flagstones (3.9), many of them with fossils, in the kitchen and elsewhere in the house (they are also seen at Sulgrave Manor and in other old houses in this part of the county). So there was certainly a quarry at that time near Byfield. None was recorded there by the contemporary historian John Bridges, but under 'Woodford' he mentioned Byfield Brook and 'in Farndon a rock of serviceable stone for building or pavements'. By Baker's time in the 1830s no quarries had been worked 'for many years' except to repair roads.

Just 5 kilometres south of Byfield, and near to Chipping Warden, is Edgcote House, built between 1747 and 1752 of excellent ashlar (3.10). Two kinds of stone are recognisable by the difference in colour,

3.8 Interior of Canons Ashby House, remodelled by Edward Dryden in 1710. The Hall floor is patterned with white limestone from Culworth and dark Marlstone Rock from Byfield.

3.9 Flagstones of dark fossiliferous ironstone from the Marlstone Rock of Byfield, in the kitchen of Canons Ashby House. Both photographs taken with the permission of The National Trust.

a darker more ferruginous type being used for the main front and garden elevations and quoins, and a somewhat lighter, rusty brown, with more calcareous particles, for the side masonry. Both types are calcareous ironstone clearly from the Marlstone Rock, with recognisable fossils, and were probably quarried on the estate: detailed building accounts, which still exist, record payment for cutting the ashlar, the only freestone purchased being Triassic

3.10 Edgcote House, designed by William Jones and completed in 1752 for Richard Chauncey, replacing the earlier house of the Chauncey family.
Below: The masonry is high-quality ashlar of calcareous ironstone from the Marlstone Rock quarried on the estate, using darker stone for the front, quoins and side window heads.

sandstone from 'Cubington' (near Leamington Spa in Warwickshire) which was for the window-and door architraves. Flagstones ('paviers') were also obtained from Byfield and Culworth. The house, according to Heward and Taylor, was built for Richard Chauncey by William Jones between 1748 and 1754; William Smith of Warwick had built the stables in 1747 but died that year.

Dark Marlstone Rock, perhaps from the Byfield area, can be seen in Northampton as pillars at the rear of the portico of All Saints' Church, built in 1701. Small blocks are also seen in the much older (Norman) priest's door at All Saints' Church, Harpole, where it was used alongside Barnack-type limestone; the Marlstone Rock here could be local. The use of Marlstone Rock indeed goes back to very early times. There are massive fonts, probably Saxon, in the churches at Kings Sutton and

Flore, which could each have been hewn from local stone. Buildings in Flore, only 10 kilometres west of Northampton in the Nene valley, are a mixture of local Marlstone Rock and imported sandstone from the Northampton Sand. The church tower appears to be Marlstone Rock (3.11), a calcareous ironstone with purplish cores, weathering to lighter brown, and in places shelly. But Flore church also has other sandy brown stone probably from the Northampton Sand. Calcareous ironstone with purplish or greenish cores, weathering brown, often with fossils, is typical of medieval churches near the Marlstone Rock outcrop in much of western Northamptonshire; St. Mary's Church, in Everdon, also has particularly good Marlstone Rock piers with fossils and sedimentary features (the windows are of dark sandstone from the Northampton Sand, also available locally).

A lighter-coloured building stone is seen in places, possibly reflecting the local geology. The old Grammar School (1600) in Daventry has light brown fossiliferous and calcareous rubblestone, from the Marlstone Formation (see 1.3). Some of the old cottages in Ashby St. Ledgers are of pale, calcareous sandy rubblestone, full of shells and belemnites. The village is built on gravels and boulder clay, but is only one kilometre from Marlstone Rock outcrops. Ashby St. Ledgers has buildings of special historical and architectural interest. The Manor House (3.12), famously associated with Robert Catesby and the Gunpowder Plot, is a fascinating complex of Tudor and Jacobean buildings with additions by Sir Edwin Lutyens in the twentieth century, as described by Christopher Hussey in *Country Life*. It is largely built of Marlstone Rock, a warm brown ferruginous limestone with a few seams of dark limonite, part rubblestone, part in dressed stone. Lutyens also used Marlstone Rock, including Hornton Stone from Oxfordshire (e.g. see 3.14) for dressings. But another type of dressed stone, as ashlar on the north side of the forecourt and elsewhere, includes calcareous sandy rock having obvious little burrows. (It was also used in church buttresses, here and at Staverton, for example).

A similar stone is seen most prominently in

Above left & left: **3.11** The fourteenth-century tower of All Saints' Church, Flore, is built of Marlstone Rock; close up, the calcareous ironstone is seen to have reddish cores surrounded by brown weathering, and thin veinlets of dark limonite (iron oxide).

3.12 The historic Manor House at Ashby St. Ledgers combines Tudor and Jacobean buildings, with twentieth-century additions by Lutyens. The south front, across the garden, bears the date 1652.
Below: A closer view of the Catesbys' Tudor façade to the west shows mellow masonry of calcareous ironstone from the Marlstone Rock Formation.

Daventry's Church of the Holy Cross, which was built by David Hiorne of Warwick between 1752 and 1758 (see 3.13). The very large ashlar blocks are of calcareous, sandy, ferruginous rock, some conspicuously burrowed; some contain bivalves and brachiopods, a few have belemnites. Towards the east end of the chancel the stone is more of a limonitic sandstone. The guide book calls it 'local ironstone', but the source is uncertain. Baker

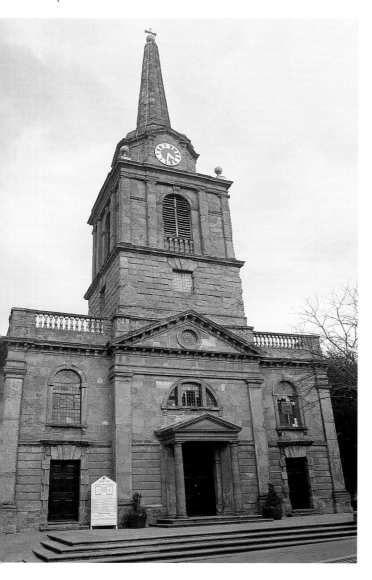

3.13 Holy Cross Church, Daventry, was built in 1752 of sandy calcareous ironstone, said to be 'local ironstone', but the source is uncertain. The ashlar blocks contain bivalves and brachiopods, and conspicuous burrows.

the Northampton Sand. These were well-known quarries, and though the distance would be about 18 kilometres, such stone could have been specified for this 'handsome edifice' (the final cost, reported by Baker, was £3486 2s 5½d).

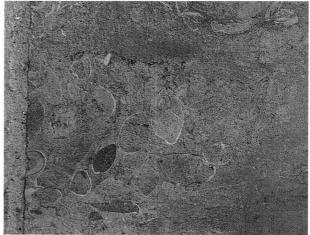

3.14 Hornton Stone, quarried in Oxfordshire from the Marlstone Rock Formation, is a calcareous ironstone commonly used for restoration of Northampton Sand buildings in Northamptonshire. It is partly greenish-grey, partly weathered brown, and fossils include brachiopods. Here seen at Welsh House, Northampton. (Hornton Stone was also used to build several twentieth-century public houses in Northampton; it is distinguished from Northampton Sand by its duller brown colour.)

mentioned quarries at Daventry 'abounding with terebratulae and remarkably fine pectenites'. A disused quarry still exists at Staverton, but the visible Marlstone Formation is a more shelly ferruginous limestone. There were quarries at Badby (see above), but the village stone, including the ashlar of the church tower, is a darker-cored calcareous ironstone. A little of the pale burrowed stone can be seen in the few stone buildings in Welton. But calcareous ironstone also occurs in the Northampton Sand (see Chapter 4), a somewhat similar stone apparently coming from the lower levels of quarries at Harlestone and New Duston (see 5.12). A pale, calcareous sandstone with burrows is seen in some of the Harlestone cottages, where it is certainly from

In villages close to Northampton (such as Flore, Harpole, and Milton Malsor), beside rubblestone cottages of local Marlstone Rock there is sandstone from the Northampton Sand, including orange-coloured ashlar of Duston sandstone. Other villages close to good (darker) sandstone from the Northampton Sand are Everdon, and Woodford Halse. Marlstone Rock villages are also not far from the western edge of the county, and rosy pink sandstone, probably from Warwickshire, makes an appearance in church windows at Chacombe and Watford, as arches under the tower at Newnham, and in the tower of the church at Crick.

The next three chapters look at more of the county's brown building stones, but these are from the Northampton Sand Formation. Superficially they may look similar to Marlstone Rock; certain rocks are indeed alike, but mostly they are distinguishable, and can be related to the geological map. The two formations are separated geologically by the 60 metres of the Whitby Mudstone Formation, and by some 9 million years.

CHAPTER FOUR

The Northampton Sand Formation (1): Ironstone as a Building Stone

Geological maps do not distinguish the two divisions of the Northampton Sand Formation (the Corby Ironstone, and the overlying Duston Member), nor can they indicate the great variety of rocks that make up the Northampton Sand. The geology summarised in Chapter 2 gives an idea of the range within this one Formation – including ironstones, sandstones, and limestones, and also rocks intermediate between them. All the Northampton Sand Formation is ferruginous to some degree, so the rocks – and the building stones – weather in warm shades of brown to light gold. But the range of rock-types varies across the outcrop shown on the map (4.1): in the eastern area (I) from the Ise valley to the Nene, and north beyond Corby, the Northampton Sand is almost entirely composed of rocks of the Ironstone division; south of Wellingborough, Ironstone continues along the fringe of the outcrop, but sandstone comes in towards the west. The western area (II) including Northampton and Duston and south-western Northamptonshire consists largely of sandstones, with a varying amount of underlying ironstone. The central area (III) has ironstone, sandstone and notably, limestones.

In this chapter we are looking at stone from the Ironstone division in the lower part of the Northampton Sand, and its use as a building stone. Long before the iron-ore industry took over the county in the 1850s, ironstone was dug for building-stone in small stone-pits and quarries on local outcrops, or under shallow overburden not far from the outcrop. In the late nineteenth and early twentieth centuries some building-stone could still be obtained, but most of the former stone-pits have been engulfed by the great iron-ore excavations and eventually lost when the landscape was restored to agriculture.

THE EASTERN AREA (1)

A. SOUTH OF KETTERING

The best examples of building in oolitic ironstone are to be seen in Wellingborough and Finedon, where it has proved surprisingly durable (see 4.2). The splendid fourteenth-century church of St Mary in Finedon, built of local ironstone (with limestone dressings of Weldon Stone) has survived remarkably well. Ironstone building continued in Finedon in later centuries: the vicarage behind the church (1688), the former Girls' Charity-School of 1712 (see 4.3) and the unusual creations of the Victorian William Mackworth-Dolben are among the interesting stone buildings in the heart of Finedon. Two stone-pits are shown on the Enclosure Award map of 1805, both in areas later worked for iron-ore.

The quality of the ironstone may be due to the nature of its weathering – which may seem contradictory. Weathering is a term covering all sorts of processes affecting rocks when they are exposed to the atmosphere, mostly detrimental and leading to crumbling and decay. The original sedimentary minerals of ironstone (ooliths of berthierine, crystals of siderite) are particularly susceptible to oxidation and hydration, forming the rusty mineral limonite; but limonite comes in various forms, some of it soft and ochreous, some of it harder, dark brown, almost crystalline, which is seen as irregular veinlets or clots, (sometimes concentric 'boxstones', as described in Chapter 2). In east and south-east Northamptonshire weathering has apparently resulted in much of the pore-space of the ironstone being filled by a form of limonite which is then resistant to further atmospheric decay; yet not so hard that the stone could not be cut and shaped. Magnified with a hand-lens the original oolitic texture can be seen, all in brown limonite (see 4.4).

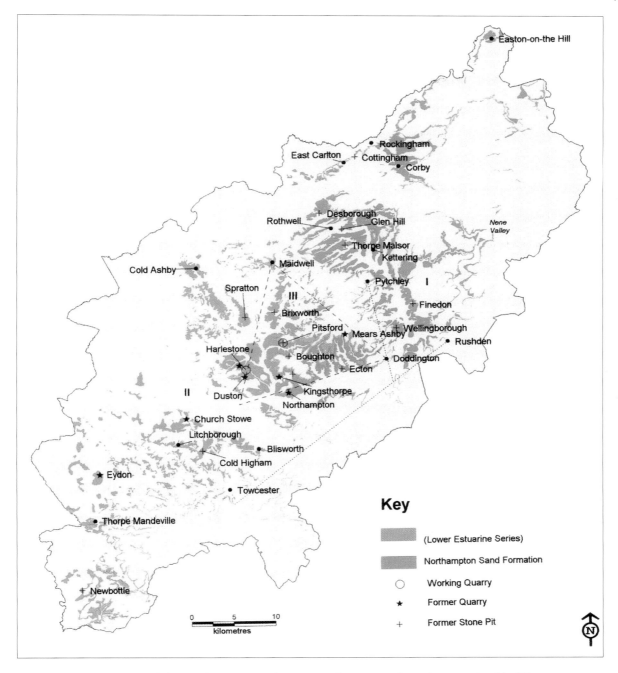

4.1 Map showing the outcrop of the Northampton Sand Formation (orange). The overlying Formation (originally mapped as 'Lower Estuarine Series') is included, to show where the Northampton Sand also generally exists under shallow overburden (boulder clay cover is not shown). But Northampton Sand dies out and is missing south-east of the line indicated from Rushden to south of Towcester; and also within 5 kilometres east of the Nene Valley to the north of Rushden. The line from Pytchley through Doddington also indicates a narrow zone where the Northampton Sand is locally thin or missing.

Three areas of Northampton Sand building stone are roughly defined in relation to the central triangle: in the area (I) to the east, north and south-east, only the Ironstone division is present; (II) in the west, south and south-west, ferruginous sandstones predominate (see Chapter 5); (III) in the central area the Northampton Sand includes varied rock-types, including limestones (see Chapter 6).

Quarries shown were at one time sources of freestone; possibly also some of the former stone-pits.

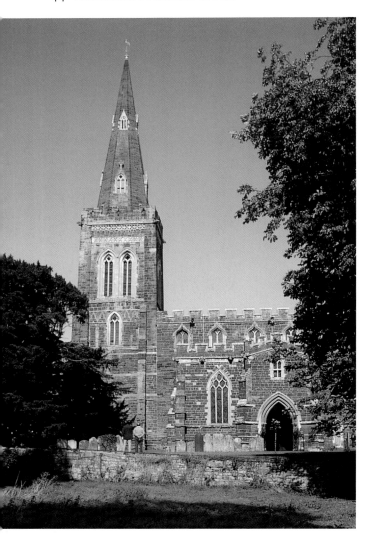

4.3 The ironstone house opposite the church in Finedon was built as a Charity-School for Girls in 1712 by Sir Gilbert Dolben. It is now a private house.

4.4 When oolitic ironstone is magnified, as with a hand-lens, the ooliths (0.1-0.2 millimetres) are recognisable. This is a limonite oolite, with a limonitic matrix, formed by the weathering (oxidation) of originally greenish sedimentary iron minerals.

4.2 Finedon Church was built in the fourteenth century from local oolitic ironstone (Northampton Sand Formation) with dressings of Weldon Stone (Lincolnshire Limestone). The ironstone masonry is composed of limonite (mixed hydrous oxides of iron); the rock was already weathered in the ground before being quarried. Despite the rusty appearance, Finedon's ironstone is surprisingly durable.

On the opposite side of the Ise valley Wellingborough occupies a south-easterly-facing hillside down to the Swanspool brook, and the Nene in the south. In the eighteenth century John Bridges observed that it was mainly 'situated on a red stone rock of which the houses are chiefly built'. Priory Cottage (1595) and the Old Grammar School (1617) built of ironstone stand by All Hallows' churchyard. They, with West End House in Oxford Street, built about 1718 (4.5) and the Manor House at Broad Green are some of the buildings that survived Wellingborough's 'Great Fire' of 1738. The Hind Hotel facing the Market Place is a gabled building of about 1645, subsequently altered in the nineteenth

4.5 A fine Georgian house, West End House in Oxford Street, Wellingborough, built of local oolitic ironstone ashlar in 1718.

century. Some of its ironstone is shelly (windows are of oolitic Lincolnshire Limestone from Ketton and Weldon). The latest – and most famous – of Wellingborough's ironstone buildings, St Mary's Church, was completed in the twentieth century, the work of Sir Ninian Comper. The interior is sumptuous. The exterior, with a simple tower (4.6), is built of excellent local oolitic ironstone ashlar; there are sporadic blocks of deep red siderite mudstone (it occurs in the top of the ironstone near here). The decorative features are of Weldon limestone.

Lincolnshire Limestone was traditionally favoured for fine work on ironstone buildings, including All Hallows in the centre of Wellingborough. But the thirteenth-century tower is striped with ironstone and local Blisworth Limestone, an example of the

4.6 Wellingborough's twentieth-century Church of St. Mary is the work of the famous architect, Sir Ninian Comper. The handsome local ironstone masonry (dressed with Weldon Stone) encloses a superb interior. The masonry is local oolitic ironstone, with sporadic blocks of red siderite mudstone (they are not burnt - this oxidised rock occurs locally at the top of the ironstone outcrop).

Left: 4.7 Windows carved in ironstone are a feature of the Church of All Saints', Wellingborough (1868). The coursed rubblestone is local Blisworth Limestone with bands of ironstone.

'polychrome banding' so common in Northamptonshire (see Chapter 11). The fifteenth-century tithe barn (once part of Croyland Abbey) is another attractive ironstone building striped with Blisworth Limestone. The Gothic-revival church of All Saints (1868) follows the tradition, but here ironstone was actually employed as freestone for the windows (4.7); the doorcase has not worn so well. There is a

4.8 Bridge Number 43 over the Grand Union Canal near Gayton is built of blocks of local dark oolitic ironstone (Northampton Sand), with a decorative band of local fossiliferous Wellingborough Limestone from the Rutland Formation.

geological explanation for the combination: the outcrop of Blisworth Limestone often occupies the hilltops, not many metres above the Northampton Sand ironstone, so both rocks were close at hand.

Wellingborough like many places has gradually covered the surrounding hillsides in twentieth-century housing estates, but from Bevan's map of 1838 and the names of fields, reproduced by Joyce and Maurice Palmer, one can see where some of the stone may have come from. There were five named Stone Pit Close, or Stone Pit Furlong; two of these were on the ironstone, totalling 30 acres, adjacent to each other along the north side of Nest Lane (an area later quarried for iron ore from about 1860). The others were on limestone on high ground north of Northampton Road, and there were several fields known as 'White Delves' where limestone was available, between the Hardwick and Kettering roads.

Ironstone is familiar in surrounding villages, such as Little Harrowden; and also as quoins, stripes or random walling in conjunction with limestone in much of east and south-east Northamptonshire (see Chapter 11). Some ironstone was used in North-ampton, though sandstones are more common (see Chapter 5), for example in the medieval Hospital of St. John, the Church of St. Giles, and a charming Regency cottage in the grounds of St. Andrew's Hospital. Ironstone was also used in a few bridges; the ancient bridge at Ditchford near Wellingborough was constructed with a coping of tough ironstone heavily knotted with limonite (its exterior ashlar masonry was of Lincolnshire Limestone from

Weldon). Further south, a pleasing bridge over the Grand Union Canal near Gayton (4.8) is of local oolitic ironstone with a band of fossiliferous limestone.

B. NORTH AND WEST OF KETTERING

Plenty of oolitic ironstone occurs to the north of Kettering, where it was widely quarried as iron ore but seldom used for building. The brown stone traditionally favoured for building in this region comes from the lowest part of the Ironstone division, below the oolitic ironstone; in the iron-ore industry it was known as 'bastard stone', which was rarely worked. It is a mixed sandy, ferruginous and calcareous rock, quite different from the oolitic building stone of Wellingborough and Finedon. It is convenient to call it calcareous ironstone, though geologists would regard it as more of a sideritic sandy limestone.

It was seldom seen in ironstone quarries because it was below the quarry floor. It has occasionally been exposed temporarily in road construction and when unweathered it is bluish or greenish grey. But where we see it most often, in buildings, blocks can have dark purplish, plum-coloured or grey cores, surrounded by a weathered zone of medium-light brown colour. Lighter coloured patches show where the sediment has been disturbed by contemporary burrowing. Pale calcareous bits can usually be seen with a hand-lens, but there are no obvious ooliths. In a few places there are bivalves or brachiopods. All this is very reminiscent of the Marlstone Rock described earlier, in Chapter 3, and indeed the rocks can look very similar. A useful distinction is that belemnites, which are often seen in the Marlstone Rock, are less common in the Northampton Sand. It also helps to remember that in northern Northamptonshire there is virtually no Marlstone Rock.

The calcareous ironstone, particularly in older buildings, is sometimes soft and crumbly. Rothwell parish church is an example (4.9). Its earliest elements take the history of working this stone back to the twelfth century, not only for masonry but for carved work – recognisable in the zigzag of the west doorway; the later stiff-leaf capitals were also carved in ironstone (on Lincolnshire limestone shafts), but some are badly worn. The ironstone blocks in the tower have dark purplish cores surrounded by paler brown weathered stone, and there are signs of

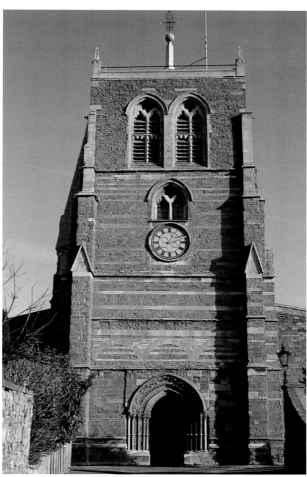

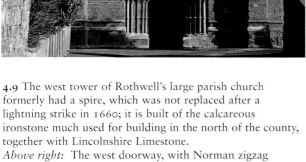

4.9 The west tower of Rothwell's large parish church formerly had a spire, which was not replaced after a lightning strike in 1660; it is built of the calcareous ironstone much used for building in the north of the county, together with Lincolnshire Limestone.
Above right: The west doorway, with Norman zigzag adapted about 1225, shows how the soft ironstone of the tower has weathered badly. The dark, purplish cores are typical of this rock, weathering to tawny brown.

burrowing but few fossils. The crumbling masonry has been patched here and there with hard limestone (this could exacerbate the trouble by directing rainwater onto the soft ironstone). Inside this great church, the ironstone remains in good condition and can be appreciated, with Pevsner, as 'singularly beautiful in its tawny colour'. Many of Rothwell's later buildings, including the eighteenth-century former Manor House, are of similar carbonate-rich stone, weathering to a medium brown, and apparently quite sound. The ironstone parts of

Tresham's Market House (1578) however have suffered badly, like the church – and have been patched with hard ginger Carstone, a Cretaceous sandstone from Norfolk.

The brown-weathering stone with purplish to grey cores is the local building stone of many villages in the surrounding area. We can see from the churches that it was worked in the medieval period – St. Giles in Desborough has a later limestone tower (see page 81) but the thirteenth-century fabric is calcareous ironstone, including the internal piers; Loddington's slender tower and carved west doorcase are all in this stone; Thorpe Malsor church has large (60 x 30 centimetre) ashlar blocks in the tower, and the spire is also apparently in brown ironstone; Draughton is another example, where some of the church stonework is twelfth century.

In all these villages the calcareous ironstone continued to be utilised in later centuries for stone dwellings, farms, and walls. Rushton, north-east of

4.10 The Georgian stable building of Rushton Hall, with ashlar masonry in local calcareous ironstone, and Weldon Stone dressings. The ironstone has dark cores, weathered to brown, and shows disturbance by contemporary burrowing. The weathering occurred before the stone was cut and set in the building - two blocks were evidently cut from the same piece of weathered stone.

Rothwell, is built of this stone, along with local Lower Lincolnshire Limestone. Rushton Hall (which was the home of Northamptonshire's famous Thomas Tresham, and built of Weldon Stone) has a fine Georgian stable-building in calcareous ironstone ashlar (4.10).

All these villages lie on the outcrop of the Northampton Sand ironstone where geologists Hollingworth and Taylor observed that 'in this neighbourhood generally, there is a thick development of the bastard stone', from two to over 4 metres. We know little about the sources of stone for the great medieval church building projects. There would have been many local stone-pits for rubblestone. Which of them could provide ashlar, mouldings, and stone piers? The villages of

Arthingworth, Braybrooke and Clipston further west, built on Lias, evidently imported the purple-cored, burrowed, calcareous ironstone (and Weldon Stone) for their churches; where was their source? Stone pits certainly existed near Desborough, Rothwell, and Loddington, and were mentioned by Morton in 1712. Thompson quoted J.R. Moore's book on Desborough (1910) which said, 'Under the ironstone is a valuable bed of building stone which has been quarried for centuries. This bed furnishes two species of stone, one known as the wheaten grain and the other as hard blue rag', but for some 50 years neither had been much used, having been replaced by brick. Morton in 1712 described pits at Glen Hill, east of Rothwell, as 'Reddish Ragg-stone, 5 Foot, in as many Strata gradually thickening from the uppermost to the lowest, whose Thickness, in that I measured, was eighteen Inches' which lay below what we would recognise as weathered ironstone. Bridges, also in the eighteenth century, recorded 'several quarries of good building stone' at Thorpe Malsor. In the twentieth century there were 'disused stone quarries' east of Faxton. No doubt more could be gleaned from old maps and documents, by the interested local historian. The stone is sometimes obtainable today from a site south-west of Geddington, owned by the Boughton Estate.

Similar calcareous ironstone is the common building stone further north around Corby and in villages alongside the escarpment overlooking the Welland valley. In some of them (Stoke Albany and Cottingham for example) the local cream-coloured Lower Lincolnshire Limestone is used as well; it is the rock that caps the escarpment. But the purple-cored, burrowed ironstone is seen from Dingley to Gretton, and in this area it can be fossiliferous, with bivalves and brachiopods. At East Carlton Church (rebuilt 1788, in a mixture of limestone and ironstone) the fossiliferous calcareous ironstone also contains a few belemnites, so that it really does look like Marlstone Rock. One of the best-known places in Northamptonshire is the picturesque village of Rockingham, leading down the steep hill to the Welland valley and the county of Rutland (4.11). Two kinds of Northampton Sand can be distinguished, the plum-cored brown, burrowed, calcareous stone which is also seen as ashlar and quoins, and more ochreous, rusty, and sandy ironstone. Both building stones are seen within the

4.11 The village of Rockingham in northern Northamptonshire, the cottages lining the main street on the steep escarpment below Rockingham Castle. They are built of ironstone (calcareous and more sandy varieties) from the Northampton Sand. Roofs are varied, including Collyweston Stone-slate and thatch.

few kilometres towards Uppingham further north. In 1875 the geologist J.W. Judd described a large quarry near Uppingham from which the lowest beds of the Northampton Sand, over 2 metres thick, yielded a hard building stone 'of a calcareous nature', beneath more ferruginous ironstone, both of which were also seen at Lyddington and Glaston. But of particular interest is Judd's detailed section of a 'great pit' below Cottingham Church, in which he recorded over 6 metres of 'hard, red rock with greenish centres (rock used for building)', succeeded by 2 metres of ironstone, and 3 metres of sands and clays which we would recognise as Grantham Formation, below the sandy base of the Lower Lincolnshire Limestone. By the 1920s Thompson wrote that they could no longer be seen.

IRONSTONE BUILDING IN THE CENTRAL AREA

This area of Northampton Sand (III on the map 4.1) has the most varied geology and building stones; the beds above the ironstone (the Duston Member) are described in Chapters 5 and 6. Ironstone, often a mixed sandy and calcareous stone, was used in rubblestone building in several villages, in Mears Ashby and Brixworth for example. At Brixworth, both oolitic and calcareous types of ironstone were also used for better building. The former stables of Brixworth Hall (eighteenth century) are built of grey-cored, burrowed calcareous ironstone; and shelly,

oolitic ironstone was used for the Workhouse in 1836. In a pit south of the Workhouse, ironstone was later quarried as iron-ore; Beeby Thompson in the 1920s recorded almost three metres of oolitic ironstone resting upon a metre of sandy, blue-hearted calcareous 'bastard stone'.

In New Duston Quarry, calcareous ironstone occurs in the top of the Ironstone beds (4.12), (below the Duston Member sandstones described in Chapter 5). The 'Rough Rag' and the underlying 'Blue Rag' are no longer exposed, but the former may be the dressed stone seen in gateways, door-cases, or quoins, and for example the pillars of the portico of Althorp stables (see 5.12). It is a calcareous, greenish cored rock, weathering light brown with limonitic veins, containing a few fossils (bivalves, brachiopods and occasionally belemnites), and with some well-defined burrows. In some ways this, too, is similar to Marlstone Rock. In central Northamptonshire it is readily accepted as a variety of Northampton Sand, but this kind of stone can be confusing when encountered elsewhere, especially in the Marlstone Rock area (see Chapter 3). The stone of the Church of the Holy Cross in Daventry, for example (see 3.13), could be from the calcareous ironstone below the sandstones at Harlestone or New Duston; it masquerades well as Marlstone Rock, even to the fossils, but the masonry grades into limonitic sandstone, more like Northampton Sand. Kislingbury Church poses a similar puzzle, but there is a clue in a stone that apparently contains the bivalve *Astarte*, a fossil found particularly in the calcareous ironstone of the Northampton Sand at New Duston.

4.12 Calcareous ironstone ('Rough Rag') from New Duston, as magnified with a hand-lens. Part of a green core is surrounded by brown weathered rock, with a veinlet of dark brown limonite. (LEIUG119508).

CHAPTER FIVE

Northampton Sand (2): the Brown Sandstones

Rich brown sandstone is perhaps the most familiar of the county's building stones. It comes from the Northampton Sand Formation, above any ironstone that may be present, or itself representing a sandy equivalent of the ironstone. It is the local variety of Northampton Sand in villages throughout its western outcrops (see map **4.1**, in Chapter 4), many of which had stone-pits providing rubblestone, and locally ashlar, in various shades of warm sienna to burnt umber; Bridges noted quarries at Adstone, Litchborough, and Winwick, and Baker also mentioned ones at Blakesley and Farthingstone. Buildings in Northamptonshire's colourful sandstones are not confined to the outcrops, as stone from the well-known quarries of Harlestone and New Duston was selected for high-quality ashlar and mullioned buildings in many other villages. Kislingbury (**5.1**), on the gravels of the Nene valley, is built almost entirely of good ginger-brown sandstone, obtainable within four or five kilometres from New Duston and Harlestone; so is much of

Harpole, and the larger houses of Milton Malsor. The sandstones are composed of quartz grains in a matrix of variably brown limonite (**5.2**).

NORTHAMPTON

The county town began as a hillside settlement between the Nene and its northern tributary, on a wide outcrop of the free-draining Northampton Sand Formation, from which innumerable springs issued at the junction with the underlying clay. The Northampton 'Sand' in fact comprises more solid rock than sand – including dense ironstone and ferruginous sandstones. With the arrival of the Normans this useful source of stone was soon recognised, and Northampton rapidly became a stone-built town, walled, with a prestigious castle,

5.1 Kislingbury village is built of sandstone of the Northampton Sand Formation, brought a few kilometres from New Duston quarries.

5.2 A thin section of ferruginous sandstone from the Northampton Sand as seen under the microscope: the clear grains of quartz, with some feldspar, are enclosed in a matrix of dark limonite. Example from Cold Higham (LEIUG122316) (field about 3 millimetres).

several churches and the buildings of many religious institutions.

Surviving remnants of medieval Northampton are few, but without doubt the finest are the two Norman churches of the Holy Sepulchre in Sheep Street and St Peter's in Marefair. The Church of the Holy Sepulchre, one of only two remaining round churches in England, was built probably by Simon de Senlis early in the twelfth century, in the form of a round nave which incorporated a circular, formerly vaulted aisle, with eight massive piers supporting an arcade; entry was by the west door, opposite the chancel built out to the east. There have been successive alterations, first by raising the Round's arcade later in the twelfth century, then by greatly increasing the size of the eastern end of the church, and the addition of a tower and spire about 1400. But the Round is still the most impressive structure. The medieval church as it grew was built entirely of Northampton Stone, probably from local quarries which have since been built over. The stone is streaky, ferruginous, with varying amount of limonitic matrix, and quite durable. A little limestone was introduced as a polychrome pattern in the Round's window arches. Extensive work by George Gilbert Scott in the nineteenth century was carried out in Harlestone sandstone, with a few bands of limestone round the apse, all in keeping with the historic building. This harmony has been sacrificed in recent twentieth-century repairs, as the round church, tower and spire, have been randomly patched with quantities of light-coloured Cotswold (Guiting) limestone (5.3). St Peter's was built, also in the twelfth century, of local sandstone and ironstone

from the Northampton Sand, but its fine architectural features are mostly in a Northamptonshire limestone (discussed in Chapter 11; see **frontispiece**).

The medieval quarries are gone, but some were apparently within the old town. Archaeologists excavating below the present Derngate site in 1981 identified a former quarry under a fifteenth-century floor. A hollow still noticeable to the north of St Giles' Church was probably another, and Beeby Thompson wrote of a former pit still remembered in the nineteenth century as 'Manning's Close' below St Giles Street, by Hazelwood Road. Other stone-pits were in the fields outside the walls.

The centre of Northampton was devastated by fire in 1675, and much had to be rebuilt, including the Church of All Saints. Some stone now came from outside the county (the Sessions House, and the pillars of All Saints' portico, are of Ketton Stone) but Morton (1712) said that most of the new buildings were constructed of stone dug from 'Northampton Field', in a freestone like that 'of Eydon and Halston, but of a lighter Colour'. Little stonework of this rebuilding in the late seventeenth century is visible, other than All Saints' Church, and this is much repaired (its north wall an amazing mixture of colours and textures from pinkish brown, streaky cross-bedded, to softer yellow-brown stone). Most of the fine sandstone buildings in the town centre – from Sheep Street through the Market Square to

5.3 The Norman Church of the Holy Sepulchre in Northampton, one of only two remaining round churches in England, was built of local brown streaky sandstone from the Northampton Sand, as were later medieval additions. Recent renovations, sadly, used Cotswold limestone.

5.4 Beethoven House in the Market Square, Northampton, built sometime after the Great Fire of 1675. The masonry is a streaky, cross-bedded ferruginous sandstone from the Northampton Sand quarried in Northampton Fields close to the town. (Beside it to the left, Welsh House survived the fire, but has been restored in Hornton Stone from the Marlstone Rock in Oxfordshire.)

George Row – are later ('Georgian'), but were probably supplied by the same stone-pits, from different stone at various levels; they vary from rust colour to clove brown (**5.4**, **5.5**). Northampton for a long time remained a compact town, with open country beyond bounds of the present Cheyne Walk, York Road and The Mounts, as we know from the map by Roper and Cole of 1807. 'Northampton Field' lay to the east of the town, towards Abington. But 'Fields' extended north of the Wellingborough Road, outside The Mounts and round to the present Barrack Road, including the area of the racecourse.

There were several quarries around the race-course in the nineteenth century, recorded by Sharp in 1870: 'on the Kettering Road, and close to the stand of the race-course, is a large pit, quarried for clay, sand, and building-stone'. This was 'a ferruginous sandstone, about 12 feet in thickness, disposed in five beds, the lower one of which is more ferruginous than those above it'. This pit, Thompson reported 50 years later, was evidently one known as Butcher's, at the back of 'Mount Pleasant', which went right up to the race-course boundary; it was subsequently built over with the houses of 'Park Crescent' (which became High Street, now Colwyn Road). The stone for St Andrew's Church (1841) and St Edmund's (1850) came from this quarry (both have since been demolished, and their stone re-used – some stone from St Edmund's went to Grundon Farmhouse in Draughton). On the north side of the race-course, in Kingsley Road, was Bass's pit recorded by Sharp, in the Northampton Sand, reaching the underlying clay. By the 1920s Beeby Thompson reported that this pit had been 'abandoned and unavailable for examination', and it was subsequently built over. He also mentioned a former stone-pit west of the racecourse at Langham Place (see Chapter 6).

Another nineteenth-century quarry described by Sharp was near St. Andrew's hospital entrance on the south side of Billing Road. It was over 6 metres deep, in ferruginous sandstone overlying fossiliferous ironstone. Several buildings in the grounds are of ironstone but the chapel (1863) is Duston-type sandstone, the hospital itself having been built in oolitic limestone from Bath in 1837.

5.5 The Judge's House, on the left, in George Row, Northampton; a Georgian building of local Northampton sandstone. Part of the Sessions House is seen to the right, built in 1678 of Ketton Stone (oolitic Lincolnshire Limestone).

HARLESTONE AND DUSTON

These two adjacent parishes north-west of Northampton were well-known sources of various building-stones – particularly sandstones, which were often called Harlestone Stone even when coming from a Duston quarry. The area of Upper and Lower Harlestone is one of the most picturesque in the county, its small groups of delightful stone cottages lying in a rural landscape around the parkland setting that is now the home of the Northampton Golf Club. The houses, big and small, thatched or tiled, are all local brown sandstone. The oldest quarries, probably medieval, remain as grassy hills and holes [SP705644]; St. Andrew's Church (**5.6**) was built entirely of local stone between 1320 and 1325. There are also remains of quarry faces behind houses. The main quarry lay south of the road to Upper Harlestone [SP705640]; it was part of the Andrews' estate which was acquired by Althorp in 1848, and continued to be worked by successive members of the Lumley family. The Lumleys are known to have been stonemasons in Harlestone for over 250 years – from the sixteenth century until the 1860s; from the first record of Thomas, who died in 1603, some 16 or 17 of his descendants, as recorded by L. Horton-Smith, were stone-cutters. They are buried in Harlestone churchyard, where there is also the fine tomb of James Whiting, from another family of stonemasons. The old Harlestone quarry was leased later in the nineteenth century by George Belton, and from 1892 by Eli Craddock, who also opened another, which is still in work on the west side of the Northampton road, south of Harlestone [SP707637] (see **2.7**). In the 1920s sandstone was being sold for making scouring bricks or 'cotters' and the lower sandstones were locally called 'Cotter Beds'; Craddock's newer quarry in Harlestone was

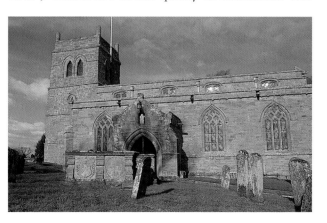

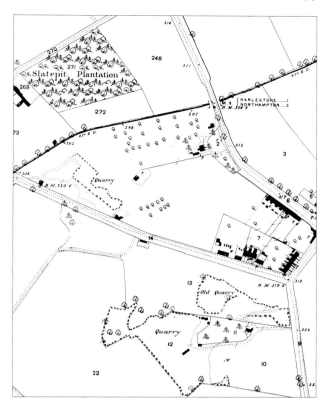

5.7 Ordnance Survey map of the New Duston quarries in 1886. Samuel Sharp in 1870 described a detailed section (see **5.9**) in the large quarried area south of Port Road which runs across the centre of the map; then worked by Samuel Golby, it later became the 'Top Pit' worked by Henry Martin (see **5.10**), part of which can still be seen at Duston Wildes. To the north of Port Road, quarrying by Golby (and later by Tennant) broke into earlier Duston slate workings ('Slate-pitt Piece' in Hewit's survey of 1692); Slatepit Plantation, shown further north, was the site of more slate workings (see Chapter 6, and **6.2**). Scale: east-west dimension is 0.5 kilometre.

Left: **5.6** St. Andrew's Church, Harlestone (1325), with the tomb-chest of master mason James Whiting who died in 1718. The tomb bears the Arms of the Masons' Company, as does the gravestone of Tubalcain Lumley (1749), one of many Lumleys nearby in the churchyard.

known to Thompson as the 'Cotter Quarry'.

The Duston quarries were at the northern edge of the parish, north and south of Port Road, in what came to be called New Duston (**5.7**). Although the area is now mostly built on, part of the largest quarry [SP713627] is accessible through the Duston Wildes housing estate (see **2.9**). A stone found inscribed with part of a date showed that this quarry complex goes back to at least the sixteenth century, and it might

5.8 Dallington Hall was built in 1720 for Sir Joseph Jekyll, Master of the Rolls, and in the nineteenth century bought by the Spencers of Althorp; it was leased to the geologist Samuel Sharp (1814-1882), who lived there for many years. It is built of Duston-type sandstone, with Lincolnshire Limestone dressings.

have been the source of the thirteenth-century sandstone lancets in St. Luke's Church in 'old' Duston. Baker suggested that 'James Whiting's stone-pit', described by Morton in 1703 as being in 'Halston', was this one, just in Duston parish. However, the actual location of the geological section listed by Morton, with the great thickness of overburden (5.7 metres, including over four metres of white sand, beneath clay and 'gravel'), overlying the various building stones (7.6 metres), is difficult to reconcile with known quarries and the geological map.

In 1870, the geologist Samuel Sharp was living in Dallington Hall which had been built, probably of Duston sandstone, in the eighteenth century (5.8). His description of the 'ancient and large' quarry at New Duston, when Samuel Golby worked it, was still applicable in the 1920s when the 'Top Pit' was worked by Messrs Henry Martin (5.9, 5.10); part of it is visible today (see Chapter 2). Sharp's section differs from Morton's, there being less than 2 metres of overburden including white sand, and a greater depth of stone (10 metres), in somewhat different divisions. Sharp evidently recorded the local quarrymen's dialect ('Roylands') for the upper 'best building-stone', otherwise known as 'Ryelands'. It is still being quarried in Harlestone, less than a kilometre to the north (see 2.8); Althorp Estate houses of 1848 in Harlestone and the Bramptons appear to be the streaky upper sandstone, in various shades of brown. The lower sandstones (below the calcareous 'Pendle') are softer and more even-textured and according to Thompson were used as

Section of Old Duston Stone-pit, giving Quarrymen's Terms.

		ft. in.	ft. in.
1.	White sand	nil to	4 0
2.	Brown soft sandstone, with vertical plant-markings ("root-perforations"?)	nil to	1 6
3.	"The Roylands"—a series of beds, each from 6 to 9 inches in thickness, very variable, sometimes hard, in which condition it is "best" building-stone, and sometimes "caly" or crumbling. These beds occur in two divisions, the building-stone of the upper being of a rich red-brown colour, and of the latter of a colder fawny-brown colour. Wood is frequently found, and I obtained from these beds a slab ripple-marked; sandy zones also occur, in which the tests of shells are perfectly preserved	9 0 to	10 0
4.	Orange sand, with rounded cores of arenaceous limestone, the remains probably of the original bed after being subjected to the action of water charged with carbonic acid	1 6 to	2 0
5.	"White Pendle"—in two beds:—		
	a. Coarsely granulated limestone, made up sometimes of oolitic grains in a matrix of calcareous cement, sometimes of crystalline angular particles with comminuted shells, more or less arenaceous in places, and containing Belemnites, large Lima nov. sp., large Hinnites abjectus, &c.	2 0 to	3 0
	b. Arenaceous and calcareous slaty beds, very like to, and called by the pitmen, "Colleyweston slate"	2 0 to	3 0
6.	"The Yellow" building-stone—consisting of six or seven beds of varying thickness, in two divisions, differing somewhat in tone of colour; these beds contain "pot-lids" of ironstone, also Cardium cognatum &c.	6 0 to	7 0
7.	"Best Brown Hard" building-stone, in three or four beds—a coarser, stronger stone than that of the other beds, but of a rich red-brown colour: it contains few fossils		6 0
8.	"Rough Rag"—a slightly calcareous sandstone, green-hearted, hard and durable, used for copings, gravestones, and building: it contains Ammonites Murchisonæ, A. opalinus, Nautilus, Ceromya bajociana, Pholadomya fidicula, Cardium cognatum, Cucullæa, &c., and a characteristic zone of Astarte elegans		3 0
9.	"Hard Blue"—a very hard blue-hearted stone, the surfaces of joints and bedding brown from oxidation: it contains the same fossils as the last bed, no. 8, excepting Ammonites Murchisonæ and the Astarte-elegans zone	3 0 to	4 0
10.	The presence of water prevents the working of the stone in this pit to a lower depth; but in an old unused pit in an adjoining field the beds for about three feet lower are exposed; and these consist of cellular ironstone, having sometimes green arenaceous, and sometimes ochreous cores		3 0

5.9 Sharp's detailed record of the quarry in New Duston in 1870. (Note: this was not 'Old Duston'.)

5.10 Duston Quarry in 1924: extracting large slabs of the 'Best Brown Hard'. (Beeby Thompson, 1928.)

5.12 The splendid stables at Althorp, the home of Earl Spencer, were built in 1733, of local Harlestone 'Redstone' from the Northampton Sand; the architect was Roger Morris.

Below: The pillars of the porticos are more calcareous dark sandy ironstone, weathering to lighter brown, with signs of contemporary burrowing, and fossils; it was also from the Northampton Sand, probably from the 'Rough Rag' occurring at a lower level in local quarries.

5.11 The Duston Yellow sandstone is recognised by its generally even texture, and bright ginger colour; it was used extensively in central Northamptonshire for ashlar and mouldings. This close view of typical masonry is at the former school (1840) in Dallington village. More streaky ginger sandstone may be from the upper sandstone ('Ryelands'). There was also a stone-pit, in sandstone perhaps similar to Duston, north of Dallington.

freestone, ginger-coloured 'Duston Yellow' or richer brown 'Harlestone Redstone'. Many buildings of the seventeenth and eighteenth centuries were built of these sandstones, which were suitable for ashlar, window-mullions and door frames (5.11). Much stone from Harlestone and Duston was used for church restoration in the nineteenth century. Some of the ashlar has weathered by spalling, and near ground level it can be badly eroded. The limonite matrix enclosing the quartz sand is not always durable .

The 'Best Brown Hard' listed by Sharp was a tougher sandstone with a cement of calcite and limonite. It was obtainable in large slabs (see 5.10), and was used for rough-dressed building stone and walling. Thompson was told by Mr Craddock that Althorp Park wall was built of it. The underlying 'Rough Rag' came from what Thompson considered to be the uppermost part of the Ironstone beds. Though sandy, it is a mixed ferruginous and calcareous carbonate rock, green-hearted but

weathering light brown, with shelly fossils (the bivalve, *Astarte*, and occasional belemnites); it is one of the calcareous rocks, often burrowed, that occur at different levels in the Northamptonshire Jurassic. The magnificent stables at Althorp designed by Roger Morris (1733) were built of typical Harlestone sandstone (5.12), the great portico pillars apparently of 'Rough Rag'.

The calcareous 'Pendle' occurring between the upper and lower sandstones at Duston, and in the Harlestone working quarry nearby, is a relatively thin development of the sandy limestones that come into the Northampton Sand further east, at Kingsthorpe and Boughton, the other side of the northern tributary of the Nene. The 'Pendle' was

5.13 Wakelyn Manor, one of the oldest houses in Eydon, with a three-storey porch and mullions of Eydon Stone, a sand stone, rich in limonite. The house was formerly thatched, and the Tudor doorway is set in an earlier round arch.

once the source of 'Duston Slates', and also building-stone. It is described in the next chapter.

EYDON

One of the best stone villages in the county, Eydon occupies a hilltop outlier of Northampton Sand. Its celebrated quarries have been infilled and restored to grass. They were mostly on the west side of the road to Byfield. The indefatigable Morton visited William Tew's pit in 1705, where nearly two metres of freestone lay below a metre of walling stone; a fine ashlar house of this date stands down the hill, beside the Byfield road. Morton observed that Eydon stone could be used for window frames and water-tables – a 'lasting Weather-Stone', better than that from Harlestone. It is very ferruginous, with sand grains set in a matrix of strong limonite. The church, with

5.14 Ashlar of Eydon sandstone from the local outcrop of Northampton Sand, at Eydon Hall, which was built for the Reverend Francis Annesley in 1789.

a fourteenth-century tower, is all Eydon Stone, including the west doorcase moulding (perhaps even the gargoyles); the blocks have knotty concentrations of limonite. South of the church are the tomb-chests of William Tew (1717), and John Wigson, carved with the arms of the Masons' Company; Eydon parish registers record many Wigsons, some of whom were masons (as were several of the Tew, or Tue, family), including 'John Wigson of Helmdon', who was buried in Eydon in 1713 'aged about 91' (see also Chapter 10).

Eydon Hall was built for the Reverend Francis Annesley in 1789, of Eydon Stone quarried on his land, and designed by the London architect, James Lewis. Marcus Binney in *Country Life* described 'the glorious colour of the local stone... each block a different shade from its neighbour... a warm brown with a hint of orange' (**5.14**). The grey sandstone pillars and windows on the garden front are a Triassic sandstone probably from Warwickshire. There are older buildings in the village, the Royal Oak with a date-stone 1692; and an interesting older house with a three-storey porch (**5.13**).

CHURCH STOWE

Finding stone for restoration these days is a problem. When Canons Ashby Church was restored in 1982, the original local sandstone, similar to that from Eydon, was no longer obtainable. Eventually, sandstone from the Northampton Sand was located in an old quarry near Church Stowe that had not been worked since 1938 [SP644575]. The light brown, fairly coarse-grained sandstones pass down through two metres of more ferruginous limonitic sandstone, overlying weathered oolitic ironstone below the quarry floor. The paler stone is poorly cemented and many blocks were rejected; the better stone below was used, though it proved a challenge to work (as Morton had said in 1712, the working of sandstone 'dulls and wears the Tools of the Stone-Cutters').

LEACHED SANDSTONES

Some buildings have a pale yellow or buff variety of sandstone with brown to orange streaks or patches. It can be seen for example in the porch of Tiffield church, and is a local stone from the Northampton Sand, affected by leaching below the white sands of the Rutland Formation.

CHAPTER SIX

Northampton Sand (3):
'Pendle' Limestones & Duston 'slates'

This chapter focuses chiefly on the central area of the Northampton Sand outcrop where the rocks overlying the Ironstone include not only sandstones but also a considerable thickness of limestone, known to Sharp, and Thompson, as 'Pendle' (pendle is a term used by quarrymen for any flaggy limestone). The area is outlined only approximately on the the map in Chapter 4 (see **4.1**, III), from Maidwell in the north to New Duston in the south-west, and across to enclose Kingsthorpe, Weston Favell, and Great Billing, reaching Mears Ashby in the east. The limestone features prominently in the building stones, and contributes to the distinctive character of villages such as Boughton, Pitsford and Moulton. In the far south-west is another area of calcareous rock in the Northampton Sand, around Newbottle.

DUSTON 'PENDLE' LIMESTONE AND DUSTON 'SLATES'

The thick New Duston sandstones have been described in Chapter 5. But sandwiched between the upper and lower sandstones here, and in Harlestone working quarry, is about 1.2 metres of pale, calcareous 'Pendle' (**6.1**), apparently the thin westerly edge of a lens of limestone in the Northampton Sand. The rock is noticeably cross-bedded, and varies from calcareous sandstone to sandy limestone with ooliths and fossil debris, and occasionally an oolitic limestone known to quarrymen as the 'Hard White'. The sandy limestones are similar to those at Kingsthorpe (see below), and were popular for building, particularly in the nineteenth century, in pale brick-size blocks. They can be seen in local terraced cottages and in some Victorian churches – those of St. Paul (I believe

now demolished) and St. Mary in Northampton were built of Duston 'Pendle' limestone from Golby's Stone-pit in New Duston.

But less well known are the rather interesting Duston 'Slates'. These are not true slates in the geological sense, as they are merely calcareous, sandy rocks that happen to split suitably along bedding planes, in fact along cross-bedding planes. The geologist Samuel Sharp in 1870 described the lower bed of the 'Pendle' at New Duston as consisting of 'arenaceous and calcareous slaty beds, very like to, and called by the pitmen, "Colleyweston slate"'; a short distance away in 'Old Slate-quarry Close', some old workings had been exposed a few years previously where the slate had been worked underground 'at some unknown distant time', in a manner similar to the 'foxing' at Collyweston (see Chapter 8). Slate was evidently dug here in the seventeenth century, as a survey of Duston by Robert Hewit in 1692 listed 'Slate-pitt Piece', more than 3

6.1 The 'Pendle' limestone, sandy and cross-bedded, is seen between the upper and lower sandstones, in the former quarry face at Duston Wildes [SP713627]. The 'Pendle' was a source of pale building stone and was mined locally for slatestone.

> JOHN LUMLEY, of Harleſton, near
> Northampton, Quarry-Man, has opened a
> Pit in Duſton Lordſhip. Any Gentleman may
> be ſupplied with Stone for all ſorts of Building;
> and likewiſe with any Quantity of exceeding
> good Slate, for ſlating Houſes, by
>   Their humble Servant,
>   JOHN LUMLEY.
> He may be ſpoke with any Market-Day at the
> Crow, in Gold-Street, Northampton.

6.2 An advertisement for Duston slates, in the *Northampton Mercury*, April 20th, 1772.

acres, and smaller 'Stone-pitts Piece' in the Heathground this side of the stream ('gutter'). In 1712, Morton mentioned Duston and Halston [Harlestone] Heaths where 'they do, or may, dig Slatestone', and an advertisement in the *Northampton Mercury* in 1772 announced that John Lumley had opened a quarry in Duston for slates (6.2). Thompson reported that clay tobacco pipes found in the workings included one dated George IV, so slates may have been worked as late as the 1820s.

The 'Old Slate-quarry Close' mentioned by Sharp was shown on his map on the south side of the stream, in the north of New Duston. The remains of shafts had been discovered when a quarry was worked for building stone on the site, north of Port Road [SP713630] (see 5.7). The shafts, according to Thompson, 'were rather narrow after being rough lined with Pendle, of moderate depth, and probably worked with a windlass, but in one case observed, there seemed to have been steps down. The area of stone got from the bottom of each shaft seems to have been rather small, hence the large number of shafts, but in one place an adit, a kind of tunnel, was encountered, which is thought by the workmen to have been about 15 yards long and some 4 feet high.' There were also similar workings on the north side of the stream in an enclosure still known as 'Slatepit Plantation' [SP712632]. Morton also mentioned that slates were 'digg'd at Pisford' [Pitsford] and 'Weston-Flavel' [Weston Favell], both places where 'Pendle' occurs in the Northampton Sand. The usual method of working slatestone required the action of winter frost to split the rock (see Chapter 8), but Morton said that the Pitsford and Weston Favell slates were suitable for use straight from the ground without frost treatment; they were presumably near-surface and already frosted.

An example of a discarded slate from the old workings is thicker than a Collyweston slate, being 15 to 20 millimetres, and it is therefore heavy; however, in the New Duston stone quarry flatter specimens of the cross-bedded, fissile calcareous

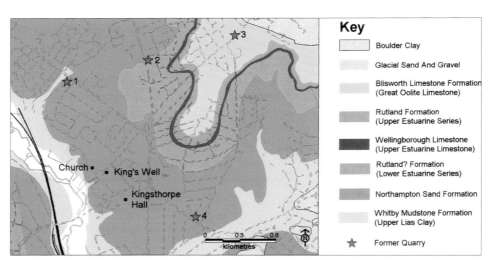

Key

Boulder Clay

Glacial Sand And Gravel

Blisworth Limestone Formation (Great Oolite Limestone)

Rutland Formation (Upper Estuarine Series)

Wellingborough Limestone (Upper Estuarine Limestone)

Rutland? Formation (Lower Estuarine Series)

Northampton Sand Formation

Whitby Mudstone Formation (Upper Lias Clay)

Former Quarry

6.3 There were quarries in several different rocks at Kingsthorpe: 1. A stone pit in the Northampton Sand sandstones, probably worked by William Bosworth [Boswell?], stone-cutter, who leased adjacent land in 1772. 2. The quarry in Kingsthorpe White Freestone, a sandstone above the Northampton Sand, mapped as 'Lower Estuarine' but probably part of the Rutland Formation. William Boswell (as he signed himself) supplied stone from this quarry for building Kingsthorpe Hall, 1773-1774 (see Chapter 7). 3. At the top of the hill were pits in Blisworth Limestone; Sharp in 1870 listed 'paving' stone for floors, but much was for lime-burning. 4. The 'Nursery' or 'Shittlewell' Pit, owned by Lady Robinson, in the Northampton Sand Formation, from which was obtained 'Kingsthorpe Pendle', a popular building stone in the nineteenth century.
IPR/37-C Brithish Geological Survey. © NERC.

6.5 Crinoidal 'Pendle' from Bradlaugh Fields 'hills and hollows', Northampton. A bed surface as seen with a hand-lens : with shell, crinoid ossicles, 'bead'-like, and also star-shaped *Pentacrinus*. (LEIUG119502).

sandstone can be as thin as 8 to 10 mm. Thin sections of Duston 'Pendle' under the microscope show very thin sedimentary layers, some more sandy, some with more slivers of shell, some with ooliths. Layering enables moisture to penetrate some planes more easily and expansion by frost causes the rock to split.

KINGSTHORPE 'PENDLE' AND SPOTTED SANDSTONE

Further east, across the northern branch of the Nene, the limestone in the Northampton Sand is substantially thicker, more than 4 metres having been measured by Samuel Sharp in 1870 in the 'Nursery or Shittlewell Pit' which lay on the hillside east of the old 'Gipsy Lane', now Kingsthorpe Grove (**6.3**, No. 4). This quarry supplied stone for the late nineteenth-century churches designed by Matthew Holding; brick-like, rock-faced masonry in pale sandy limestone is characteristic. St Matthew's, built of Kingsthorpe and New Duston 'Pendle' (with Bath Stone dressings) is a fine example (**6.4**). The blocks were cut and laid with the main bedding, but they show oblique cross-bedding and contain fossil debris, including the small bead-like remains of crinoid stems ('sea lilies', a type of echinoderm living attached to the sea-floor), some being the star-shaped *Pentacrinus* (**6.5**). Some brown limonite occurs, dispersed and in veinlets in typical 'Pendle'.

The Kingsthorpe quarry was infilled during the 1920s, and houses were later built on this hillside. There is still a pleasant old wall of this local stone along the east side of Kingsthorpe Grove. It is

6.4 St. Matthew's, Northampton's foremost Victorian church, designed by Matthew Holding and built in 1891-1894 in 'Pendle' limestone from Kingsthorpe and Duston, with Bath Stone dressings. The masonry is composed of rock-faced blocks of sandy limestone, slightly ferruginous, cross-bedded, and containing fossil crinoid debris.

6.6 The Roman Catholic Cathedral House and St. Andrew's Chapel in Northampton were built outside the town in 1825, probably from a stone-pit on the site of the present Langham Place. The masonry is distinctive oolitic limestone ashlar, strongly cross-bedded and sandy, from the Northampton Sand.

common in the walls of the old village of Kingsthorpe, along with examples of several other interesting Kingsthorpe stones. One is a conspicuously spotted rock which is largely peculiar to Kingsthorpe (though something similar was once seen in New Duston quarry). It is a brown sandstone with pale yellow or white blotches. These are concretionary nodules, cemented by calcite. Their irregular shape suggests the sands were probably burrowed, and calcium carbonate present in groundwater became precipitated in the burrows. This spotted rock can be seen in walls in the old village, especially by the south side of the High Street, where there appear to have been quarries. The spotted sandstone is exposed *in situ* below a wall near the corner of Kingswell Road and The Green. It occurs in the lower part of the Northampton Sand, some 2 to 3 metres above the King's Well spring that bubbles out near the junction with the underlying clay. Another 'peculiar' rock recorded by Sharp in the Kingsthorpe Nursery quarry was intricately and closely layered in thin ribs, and Thompson mentioned one with closely alternating ferruginous layers; something like it forms the bases of the buttresses at St. Peter's Church (see **Frontispiece**).

'Pendle' is not found in central Northampton, but calcareous rock apparently came into Bass's pit, just half a kilometre south of the Nursery Pit, as seen by Sharp; Thompson thought the first 'Pendle' houses were built by Mr Bass in Kingsley Road. He also reported that St. Katherine's Church had reputedly been built (1838) from a stone-pit on the site of Langham Place, north-west of the racecourse; this church has now been demolished, but Thompson

saw that it was a light-coloured, 'strongly false-bedded' stone which he regarded as 'hybrid' between Northampton sandstone and the Kingsthorpe 'Pendle'. The Roman Catholic Cathedral House, with St. Andew's Chapel (both built in 1825), almost opposite Langham Place, are probably examples of this stone. The long ashlar blocks are oolitic limestone, conspicuously streaky and cross-bedded, with sandier layers; the blocks are larger than usual Kingsthorpe 'Pendle', and crinoids are not as common (**6.6**). E.W. Pugin's rock-faced Roman Catholic Cathedral of 1864 is sandy, cross-bedded oolite, but by then perhaps came from either Bass's, or the Nursery pit.

HILLS AND HOLLOWS, ABINGTON AND WESTON FAVELL

East of the Nursery Pit, towards the Kettering Road, is an extensive area of 'hills and hollows' which are the remains of old quarries in 'Pendle' limestone, with some brown sandstone above. In 1895 the Northampton Golf Club took over the site, inheriting a landscape that lent itself admirably to the purpose. A hundred years later they moved away, and it has become a Local Nature Reserve known as Bradlaugh Fields. No-one knows when these quarries were worked, even Thompson, whose knowledge would go back into the nineteenth century. The quarries must be older, though not necessarily as old as the Roman settlement on the site of the present supermarket. There are no buildings of 'Pendle' here. The site of medieval Moulton Park was less than a kilometre to the north, and surrounded by a stone wall which, according to records, was often need of repair – but the hills and hollows cover a bigger area than mere stone-pits for a boundary wall. The quarries were alongside part of the former parish of Abington which stretched from north of the Kettering Road down to the Nene. Little remains, in Abington Park, of the old village of Abington that existed in 1671 (on a map shown by E.E.Field), other than the church, the former manor house (now Abington Museum), both of which have undergone reconstruction, and a few relocated, rebuilt cottages. All are built mainly of 'Pendle' limestone. The Abington manor house of the fourteenth century (the residence of the then High Sheriff, who was also Keeper of Moulton Park) was rebuilt in 1500, and again in 1678 when it had been acquired by the Thursby family; but the south and east fronts seen today were rebuilt in 1740, by the well-known architect, Francis Smith of Warwick (6.7). The masonry is pale sandy limestone with limonite veinlets, containing shell and crinoid debris. The quoins and doorcase are calcareous ironstone, sandy greenish-cored Northampton Sand (similar to New Duston's 'Rough Rag'). There could have been other sources of 'Pendle' for Abington village, north of the Wellingborough Road, but there was a convenient route for stone to be carted the 1.5 kilometres from these obvious quarries, down the old 'Plum Lane' (near the present Park Avenue North).

Towards Weston Favell the Northampton Sand is mostly 'Pendle', a pale sandy limestone, and the stone walls of Abington and Weston Favell are typically made of flat blocks of it. In 1923 Thompson measured some 3 metres in Quinton's quarry where the Cherry Orchard school playing

field is now. There is still some rock exposed in the base of the wall built over it. Sandy 'Pendle' is also the main building stone of Great Billing.

BOUGHTON, PITSFORD AND MOULTON

Boughton has an old quarry by the Green, outside the present village (a large modern quarry west of the village has now closed). Its cottages are of 'Pendle' rubblestone in flat blocks of pale brown calcareous, sandy stone (see 6.8). Boughton Park has a long boundary wall beside the main A508 road to the west, where William Wentworth built the Hawking Tower (c.1756), originally as a lodge to Boughton Hall, of local stone.

The great lens of 'Pendle' limestone in the Northampton Sand is over 5 metres thick in a quarry which is currently worked, mainly for crushed stone and walling, south of Pitsford village [SP757671]. It includes hard limestone, blue-hearted before weathering to buff. Morton mentioned a stone pit

6.7 Abington Museum in Abington Park, Northampton, a former Manor House, was remodelled in 1740, when the south and east fronts were designed for John Harvey Thursby by the architect Francis Smith of Warwick. The masonry is ashlar of sandy 'Pendle' limestone with crinoids (possibly from the 'hills and hollows' by the Kettering Road?) The quoins are darker calcareous ironstone similar to the 'Rough Rag' at New Duston.

6.8 Cottages in Boughton, built of rubblestone in flat blocks typical of the local calcareous, sandy 'Pendle' of the Northampton Sand.

with 'broad Stone that rises, as the Diggers express it, with a Head and a Bed, that is an eaven Side or Edge, and an eaven Surface: Insomuch that they lay pretty handsome Floors of it'. The old village is an attractive assemblage of rubblestone cottages, many having flat blocks of the sandy limestone. Pitsford Hall is now occupied by Northamptonshire Grammar School; built originally for Colonel James

6.9 Pitsford Hall, designed by John Johnson (*c.*1775), was built of local sandy, crinoidal 'Pendle' limestone from the Northampton Sand; the eastern building, seen here, is decorated with grey Kingsthorpe Sandstone (see Chapter 7).

Money about 1775 it was, in Baker's words, 'a handsome house' flanked by separate pedimented service buildings, one of which remains. The architect, John Johnson, was also responsible for Kingsthorpe Hall (see Chapter 7). The Hall masonry is well-dressed local golden 'Pendle', and the east building the same but rubblestone, with dressings of Kingsthorpe Sandstone (see Chapter 7) (**6.9**).

Field walls towards Moulton, (and around Overstone and Sywell, noticeably the Overstone Park wall along Ecton Lane) are all made of flat 'Pendle' stone. Moulton village has many old buildings of it, including the church. The 'Pendle' beds continue to the north, but in Brixworth more use is made of local ironstone and brown sandstone, and by Maidwell the golden 'Pendle' is accompanied by sandstone and also Lincolnshire Limestone.

MEARS ASHBY

At Mears Ashby, the calcareous rock is less flaggy, and more thickly bedded, overlying ironstone, and beneath beds of brown sandstone. The limestone here made a good freestone, though it was apparently only about a metre thick. There were two quarries in the parish, both worked in Morton's day, around 1700: 'One of the Strata in the Quarry called High-Delves at Mears Ashby has the Name of Freestone, by way of Excellency, as it does not fret a

6.10 Mears Ashby Hall, built in 1637 probably for Thomas Clendon, in Mears Ashby sandy crinoidal limestone. The wing to the right was added in 1860 in the same stone. The Mears Ashby limestone was a freestone, and large blocks were carved for the porch. (The stone can be examined by the gate.)

Tool, like their ordinary coarser Sandstone . . .'Tis a Weather-stone: And is wrought into handsome Ashler, Window-Cases, Piers &c.'. High Delves was beside the Hardwick-Sywell road, a site still visible as wooded 'hills and holes' [SP844685]; Hundhill quarry was closer to the west side of the village [SP833668], and was infilled in 1980. Morton gave a detailed list of the beds he saw in High Delves Quarry, the 'Freestone Rag' and the 'Gray Freestone' being the calcareous rock we now call Mears Ashby Stone. The stone is not at all 'gray', but a light-coloured yellowish gold. Thompson called it Yellow Building Stone. It is distinctly cross-bedded, with quartz sand grains having an oolitic coating, calcareous particles, including many crinoid ossicles, and a variable amount of streaky limonite. (It is one of the few varieties of Northampton Sand that is appropriately repaired with yellow Guiting limestone.)

Mears Ashby Hall was built of the sandy limestone in 1637, and extended in similar stone in 1860 (6.10). Other ashlar buildings in this very pleasant village include the early eighteenth-century

6.11 The exquisite fourteenth-century tower and spire of St. Mary's, Wilby, is an early example of Mears Ashby Stone. *Above:* The west portal moulding showing the golden sandy limestone cross-bedded, and containing fossil crinoids.

Manor House, and the Callis family farmhouse of 1848. But the local Northampton Sand comprises ironstone (sandy and calcareous) beneath the celebrated limestone beds, and a durable dark brown limonitic sandstone above, also used for building; all these varieties appear in the rubblestone cottages and in the medieval church, but selected Mears Ashby Stone apparently only came in with the fifteenth-century clerestory. However, Mears Ashby Stone had been used earlier – and at its most decorative – in the fourteenth-century tower and spire of the church at Wilby (**6.11**).

Several of the mansions in the surrounding villages appear to have been built or faced with this stone, Sywell Hall (*c.*1617) just a kilometre distant, Orlingbury Hall (1709), and Ecton Hall (1756), both within 5 kilometres of the quarries. Further afield it can be recognised in the seventeenth-century part of Delapré Abbey in Northampton, added by Zouche Tate. The ashlar blocks are larger than the 'Pendle' at Abington or Pitsford Hall. The stone is particularly well seen in the Church of St. Mary in Orlingbury, which was built in 1843 (**6.12**). The blocks are large, some nearly a metre in length, cross-bedded, with ferruginous streaks and abundant crinoid ossicles.

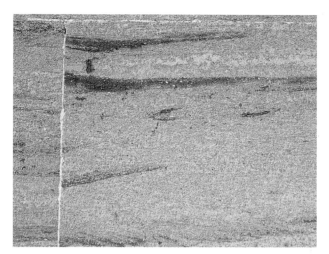

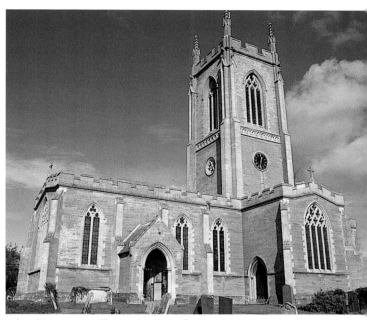

6.12 St. Mary's Church, Orlingbury (*right*), was built of Mears Ashby Stone in 1843, with dressings of oolitic Weldon Stone (Lincolnshire Limestone). *Above:* The ashlar is in large blocks of cross-bedded sandy limestone, with ferruginous (limonitic) streaks, and plenty of crinoid ossicles.

NEWBOTTLE AND THORPE MANDEVILLE

Calcareous rocks also occur in the Northampton Sand in the south-west of the county. Up to 1.5 metres of variably sandy and oolitic limestone can be seen old quarries in Newbottle Spinney [SP517365], and similar rock was probably quarried west of Charlton, a village built of a mixture of the Northampton Sand and local Taynton Limestone Formation. Thorpe Mandeville rubblestone is a pale brown, sandy or pebbly calcareous rock with crinoid fragments; Aveline and Trench in 1860 reported that building-stone had been quarried at Thorpe Hill Farm.

CHAPTER SEVEN

Kingsthorpe White Sandstone

In the eighteenth century a new stone was introduced into Northamptonshire architecture, and remained popular for about a hundred years. It was a white or pale grey sandstone that came from a quarry on the north side of Kingsthorpe. The historian George Baker observed (1823-30) that 'to the east of the Harborough road is a quarry of White Freestone of a sandy texture with a remarkably fine grit and a tendency to harden on exposure to the air. The General Infirmary and Barracks at Northampton and some of the neighbouring seats were built of this quarry which is of considerable local celebrity'. The quarry still existed, but apparently was no longer working in Samuel Sharp's time (1870), and is shown on the Ordnance Survey map of 1886, situated beside the Harborough road about where Chalcombe Avenue now is [SP751641] (see **6.3, No. 2**). The geology is discussed later, when it will be appreciated that there is good reason for this chapter to have come later in the book (before Chapter 10); on balance it seemed appropriate to maintain the geographical association with Kingsthorpe established in the previous chapter (this

building stone is not known to occur anywhere else).

In his history of the Northampton General Hospital, F.F. Waddy described the new Infirmary building erected in 1793 on 'Northampton Fields' just outside the town to the east (**7.1**), for which Mr Benjamin Drayton, then owner of the Kingsthorpe quarry, gave £1000-worth of white sandstone. The original front elevation was subsequently lost to view when in 1872 the hospital was extended forward in Bath Stone, and most of the rest has been replaced. The Porter's Lodge and wall on the corner of Cheyne Walk, built of Kingsthorpe Sandstone in 1841, was only demolished in the late twentieth century. The cavalry barracks built in 1796 on the north side of Northampton was another large complex which included ashlar of Kingsthorpe Sandstone, but no stone is visible now in the buildings remaining beside

7.1 The Infirmary (Northampton General Hospital), as built of Kingsthorpe Stone in 1793. The hospital has since been covered in Bath Stone. Coloured engraving from a water colour by W.W.Wells (1762-1836). (Northamptonshire Libraries and Information Service.)

the modern Post Office sorting office. Kingsthorpe Sandstone has therefore almost disappeared from central Northampton, despite its use for prestigious buildings. A small remnant of grey Kingsthorpe Stone masonry can still be seen above a shop in The Drapery, which once used to be Jeyes the chemist. But there are interesting examples remaining in Kingsthorpe and some of the villages.

The stone may have been first quarried in the 1760s, when it was used by William Wentworth, second Earl Strafford, of Boughton Hall (3 kilometres north of Kingsthorpe) to build an obelisk, sometime after 1764. The obelisk, an imposing Egyptian-style monument 30 metres high (**7.2**), was built on top of a hill to the south of Boughton Park. But where once it was a focus to draw the eye, it is now surrounded by housing and quite difficult to find. Simon Scott (in *The Follies of Boughton Park*) explains how Wentworth came to erect this impressive monument in memory of his friend William Cavendish, fourth Duke of Devonshire, a respected national figure, for a short time Prime Minister, who died in 1764 at the age of 45. The sandstone at the base is in large blocks up to 1 metre wide, 0.4 high and 0.4 deep, a pale creamy grey colour, with dark grey slivers of possibly carbonaceous, finer silt. The blocks have hardened at the surface, but the stone is nevertheless soft and easily scratched. Above the base the blocks are smaller, up to 0.6 by 0.3 metres, slightly yellowish, and less weathered. Repairs have been carried out in another white sandstone, rather coarser grained, from outside the county.

The quarry does not appear on Kingsthorpe's Enclosure Award map of 1767 (NRO), but by 1773 William Boswell was supplying white sandstone from here for Kingsthorpe Hall, one kilometre to the south (**7.3**). This mansion, built for James Fremeaux, was designed by the Leicester-born architect, John Johnson, who had by now made his name in London. The detailed building accounts for Kingsthorpe Hall are quoted in Nancy Briggs' biography of Johnson. He was responsible for several houses built for the Northamptonshire gentry about this time, including Carlton Hall for Sir John Palmer (since rebuilt, in 1870), and Pitsford Hall for Colonel James Money (now Northamptonshire Grammar School).

Kingsthorpe Hall, a fine ashlar building, was eventually purchased from F.H. Thornton by

7.2 The Obelisk in Kingsthorpe, 30 metres high, built by William Wentworth of Boughton Hall sometime after 1764, of blocks of Kingsthorpe White Sandstone [SP753652].

Northampton Borough Council and became a Community Centre, surrounded by Kingsthorpe Park; but it has suffered recent disaster. When high levels of radon gas were found in the building in 2000, it was immediately closed (the radioactive gas is known to emanate from underlying Northampton Sand, and special ventilation is often required to prevent its accumulation in buildings). Since then, there has been a damaging fire in the roof, which is currently covered in a polythene tent, and windows are boarded up; but it is due to undergo renovation. Kingsthorpe Hall is of more than architectural interest. The stone itself tells of an episode in the geological history of Northamptonshire when the

7.3 Kingsthorpe Hall, built by the architect John Johnson, 1773-5, in Kingsthorpe White Sandstone ashlar, stands in Kingsthorpe Park [SP750628]. The sandstone of Kingsthorpe Hall (*right*) is an interesting example of 'geology above ground': the stone has dark vertical traces of the roots of Jurassic marsh plants. (It is a Regionally Important Geological Site.)

Jurassic sea lay further south, and the Midlands became a broad expanse of marshy terrain occupied by horsetail plants. The remains of roots of these plants can be seen as dark carbonaceous vertical traces in the pale grey stone on the south side of the building (see detail, **7.3**). The quarry has long gone, but this building provides what, hopefully, will once more be an accessible example of 'geology above ground'.

A fine eighteenth-century mansion, privately owned and not open to public view, is Little Houghton House (**7.4**), which has architectural features identified as again the work of John

7.4 Little Houghton House, a gracious residence (*c.*1780) with some of the hallmarks of the architect John Johnson, is faced in Kingsthorpe White Sandstone ashlar. Alterations were carried out in the same stone in the nineteenth century. More recent repairs in Weldon limestone blend well.

Johnson in about 1780. The building is faced with Kingsthorpe Stone, a smooth ashlar of pale creamy grey fine-grained sandstone, some blocks having small lenticular slivers of fine silt, others with hollow tube-like traces of plant rootlets. When the main entrance was moved in 1851, and certain alterations were made to windows, the Kingsthorpe quarry was apparently still able to supply stone for window mouldings and the new porch. Repairs in recent years have been carried out in limestone, Weldon Stone, reclaimed from Horton House when it was demolished; the two stones blend well in colour and, being similarly porous, appear to be compatible.

Brixworth Hall, built in the 1740s and demolished in 1954, was described as 'yellow sandstone'; pictorial records, including Perrin's watercolour in Northampton Central Library, suggest it was a light-coloured rock from the Northampton Sand. The later eighteenth-century orangery, which has a façade of grey Kingsthorpe Sandstone, was moved from Brixworth to Kelmarsh Hall (itself a handsome brick building), where it can be seen in the garden. Spratton Hall, across the valley from Brixworth, is a former country house (now a preparatory school) which was built of Kingsthorpe Stone sometime before 1778. A one-time pupil (an eminent geologist) observed that the stone was easily scratched, and small boys were able to make their mark. The porch and later repairs are in Bath Stone.

Pitsford Hall, designed by John Johnson about 1775, has been mentioned in Chapter 6; it was built of golden calcareous 'Pendle' from the local Northampton Sand. A little Kingsthorpe Sandstone was used for the decorative band and gable moulding of the separate east building (see **6.9**), but some has been replaced in other stone. East Haddon Hall, built (1780-83) by John Wagstaff of Daventry, which has been compared with John Johnson's buildings, is constructed not of Kingsthorpe Sandstone, but of a Triassic sandstone from Attleborough, Nuneaton.

Not many churches were built in this period. But Overstone's church of St Nicholas was constructed in 1807 of Kingsthorpe Stone; including some of the mouldings. The grey sandstone has the usual little streaks of silty clay. There is also later Weldon Stone (and a new addition in Bath Stone), but yellow Guiting Stone for repairs is an unfortunate choice, not only on account of its colour but also its

7.5 St Nicholas Church, Overstone, was built in 1807, of Kingsthorpe Sandstone; the later window (right) is Weldon Stone (Lincolnshire Limestone). More recent repairs to worn masonry are in yellow Cotswold limestone from Guiting.

hardness and contrasting porosity (**7.5**). Rain running off the new stone will, I fear, add to the decay of the soft sandstone.

Sharp described the Kingsthorpe quarry as exposing clay in the top, which was overlying nearly 4 metres of white or grey sand, stratified, ferruginous in places, with bands and patches of clay, 'more or less coherent and sometimes so indurated as to make a very durable building-stone'. Near its base was a 'plant-bed', a band 15 to 30 centimetres thick, of sand laminated with vegetable matter, which overlay a bed with numerous root-perforations, penetrating to a depth of from 30 to 60 centimetres. Below this bed was dark brown sandstone which he recognised as the 'variable' beds of the Northampton Sand. On Sharp's evidence, the white or grey freestone came from indurated sands overlying the Northampton Sand: the white sandstone with dark silty streaks

7.6 This modest nineteenth-century cottage of Kingsthorpe White Sandstone, just 300 metres from the former quarry, is one of the last remaining in Kingsthorpe.

coming from the beds above the 'plant-bed', and the sandstone with root markings as seen at Kingsthorpe Hall, from just below the 'plant-bed'. The white sands overlying the Northampton Sand were mapped by the Geological Survey as 'Lower Estuarine Series' – which is generally renamed the Grantham Formation; but recent research suggests that in this part of the county (see Chapter 2), these may in fact be the deposits of the later Rutland Formation, resting unconformably on the eroded surface of the Northampton Sand.

Very little of the white sandstone remains in Kingsthorpe. The row of cottages photographed by Thompson in the 1920s seems to have gone, and the bold quoins edging some brown sandstone cottages in the village High Street have mostly been replaced. There is just one white sandstone terraced cottage on the Harborough road (7.6), and part of a stone cottage opposite. A little of the stone is recognisable in other villages – in several buildings in Pitsford, and as a cottage façade in Holcot. But the 'celebrated' stone has become a rarity.

CHAPTER EIGHT

The Lincolnshire Limestone [1]: *with Collyweston Stone Slates*

The Lincolnshire Limestone has for centuries provided some of the country's finest building stones. Three great cathedrals of eastern England are built of it – Lincoln, Ely, and Peterborough, as are many of the historic stone colleges of Cambridge. From Lincoln Cliff, where it is 30 metres thick, the limestone can be traced by a string of well-known quarries southwards through Ancaster, Clipsham, Casterton, Stamford and Ketton. The limestone becomes thinner as it noses into Northamptonshire, disappearing altogether by Kettering, and just reaching Maidwell in the west (see **2.14**); hence it is confined geologically to the north of the county (**8.1**). Here were the historic quarries of King's Cliffe, Weldon and Stanion. The famous Barnack quarries were in the adjacent Soke of Peterborough, formerly part of Northamptonshire. The well-known freestones (which are described in the next chapter) came from the Upper Lincolnshire Limestone, which in Northamptonshire occurs in only a few localities within the wider area of Lower Lincolnshire

Limestone. This chapter looks at the Lower Lincolnshire Limestone, which is the local building stone of more than 20 villages on or close to the outcrop, and is the source of Northamptonshire's Collyweston 'Slates'.

The Lincolnshire Limestone forms a broad plateau in northern Northamptonshire from high above the Welland valley at Easton-on-the-Hill, Collyweston and Duddington, across to Wittering, Barnack, and Ufford in the former Soke of Peterborough, where it is cut by rivers running south to the Nene valley at Wansford. From Collyweston and Duddington the outcrop continues south-westwards along the Welland escarpment through Wakerley and Gretton, above the villages of Rockingham and Cottingham, as far as Wilbarston. South of Duddington, between the Welland to the west and the Nene in the east the Lincolnshire Limestone is covered by later Jurassic strata and a progressively more extensive blanket of glacial boulder clay. Three of the Nene's tributaries have cut quite deep valleys into this landscape,

8.1 Map of the outcrop of Lincolnshire Limestone in northern Northamptonshire. The overlying Rutland Formation is included to show where the limestone may occur beneath it; but note that the limestone dies out along a line from about Fotheringhay and Tansor (south of Nassington) to Maidwell, and is absent south-east of this line. (The limestone seen in the south-eastern area is Blisworth Limestone, see Chapter 11).

The following villages, on or close to the outcrops, are built of Lincolnshire Limestone, (those in brackets having significant other local stone): Apethorpe, Blatherwycke, (Brigstock), Bulwick, Collyweston, Cottingham, Deene, (Deenethorpe), Duddington, East Carlton, Easton-on-the-Hill, (Fotheringhay), Geddington, Great Oakley, Gretton, Harrington, Harringworth, King's Cliffe, Laxton, Little Oakley, (Maidwell), Nassington, Newton, Pipewell, (Rushton), Stanion, Wakerley, Weekley, Weldon, (Wilbarston), (Woodnewton), Wothorpe, Yarwell.

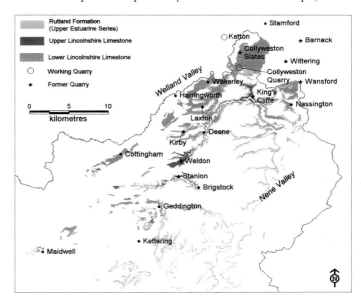

8.2 The delightful village of Duddington, like many in northern Northamptonshire, has roofs of Collyweston Stone Slate.

exposing Lincolnshire Limestone in many places along the way: the Willow Brook dissects it from Weldon near Corby, north-east to Bulwick and Blatherwycke, and round to King's Cliffe and Woodnewton (where Lincolnshire Limestone dies out); the Harper's Brook cuts into it intermittently from Pipewell to Little Oakley and Stanion, as far as Brigstock (where it is cut off by a fault); and along the upper Ise valley it is exposed between Rushton and Geddington, but here the limestone is much thinner, and it has disappeared from the geological succession by Weekley, near Kettering. Lincolnshire Limestone is missing from the geological succession to the south-east (see Chapter 2); it just reaches the Nene Valley near Fotheringhay and again near Nassington, widening at Yarwell and Wansford, and tapering away by Castor and Peterborough in the east.

COLLYWESTON STONE SLATES

The county's famous stone slates come from the area of Collyweston, Easton-on-the-Hill and Wothorpe; some were also obtained for a time at Kirby, near Deene, and locally at Harringworth and Wakerley, all places along a narrow NE-SW belt and within two to three kilometres of the east side of the Welland valley. Geologically they come from the lowest beds of the Lower Lincolnshire Limestone in the few localities where these beds form a sandy limestone with the necessary properties to split into thin plates; in most other places the lowest beds are sandy limestones that are not suitably fissile, or are only partially cemented sands. Here the sandy limestone is broadly cross-bedded, with fine layers at a low angle to the main bedding, containing tiny mica flakes and some shell, the rock being cemented by calcite to form a coherent bed. The bed can be from a few centimetres to about a metre in thickness, the lower surface rounded and undulating, and resting on soft, uncemented sands near the junction with the underlying Grantham Formation.

When the damp rock is exposed to frost it splits naturally into thin layers. Material suitable for roofing would have been first found where it lay at the surface. In due course the slatestone was obtained by digging into the limestone plateau around Collyweston and Easton-on-the-Hill and an industry developed which was able to supply, for instance, 14,000 slates for Rockingham Castle in the fourteenth century and many subsequent consignments for major buildings in the county as well as for several of the Cambridge colleges, conveyed via the fenland waterways. Collyweston-slated roofs are an attractive feature of many villages in northern Northamptonshire (8.2, 8.3).

8.3 The stone slates grade from the smallest at the ridge to the largest at the eaves.

(*Right*) **8.4** The underground gallery in a Collyweston 'slate' mine was just deep enough for a miner to lie on his side, using a foxing pick to loosen the slate bed above. The roof was supported by pillars of stone.

The industry which thrived for centuries developed its own particular skills. Shafts had to be dug, sometimes 8 or 10 metres through the thickness of Lower Lincolnshire Limestone to reach the soft sand underneath the Collyweston horizon. From the shafts galleries were dug horizontally in the sand just high enough for the miner to lie on his side, using a foxing pick to undermine the slate bed (**8.4**). He supported it at intervals with stone props, but he would listen for clicks in the bed (known as 'talking'), indicating that the bed was about to fall, when he would withdraw, removing the props. Working sections of about 4 metres were determined by natural joints. The work was done by candle-light, or later, by paraffin lamp. The lumps of stone (slate 'log') were then raised to the surface, laid out and constantly kept damp, waiting for a succession of frosts to split the stone. Slate mining was therefore mostly carried out in December and January.

Later in the year the slaters could tap open the log with a cliving hammer, and begin the skilled work of dressing the slates in a range of sizes from the 6-inch 'mope' for the top of a roof increasing to the 24-inch 'long ten', sometimes larger, for the eaves (**8.5**). The slater's 'heap' or 'thousand', consisting of 840 slates and 13 large ones, for 18 square metres of roofing, weighs about a ton.

Slate mining used to be done by individual copyholders, and there were many shafts and a warren of underground galleries, but the industry

8.5 David Ellis, master slater, pecking the hole in a trimmed stone slate.

8.6 Collyweston stone slates being fixed to the roof of the Guildhall in the City of London by Andrew Stubbs in 1998.

has always been very labour-intensive and by the 1960s (Purcell reported) only three mines were still operating. A succession of mild winters has hastened the decline. Underground mining has now ceased altogether, but slate 'log' is obtainable from the bottom of the opencast Collyweston Quarry, though still requiring winter frost. The Collyweston Stone Slaters' Trust was formed in 1982 to support the declining industry. An experiment was carried out in the 1980s to freeze the log artificially, but the practice has not been developed on any scale. However, modern cutting equipment is now used to prepare the slate, and new Collyweston slates were used in 1998 at the Guildhall in the City of London (8.6). Once prepared in the traditional way the slates are extremely durable, which means that slates salvaged from demolished buildings can be re-used, to meet the demand for this traditional natural stone roofing.

Around Wittering (Cambridgeshire) the equivalent beds once provided slabs, about 5 centimetres thick, of tough sandstone known as Wittering Pendle, which were used for steps and flooring; small blocks were also set on edge for stable yards (as can be seen at Burghley House, Stamford).

BUILDING STONE FROM THE LOWER LINCOLNSHIRE LIMESTONE

Well-trimmed cream rubblestone is typical of villages on or close to the outcrop. Where the feather-edge of the limestone forms outliers at Maidwell and Harrington small blocks of fine-grained cream limestone appear as rubblestone in the village walls, cottages and the churches, at Maidwell interspersed with more golden, sandy Northampton Sand 'Pendle' and brown ironstone. This limestone lies only a few metres above the Northampton Sand, so

8.7 A cottage of Lower Lincolnshire Limestone in Harringworth, a village close to the outcrop.

Above: **8.9** Lower Lincolnshire Limestone rubblestone: a medium-fine grained sandy limestone, in the Priest's House (early sixteenth-century) at Easton-on-the-Hill. (National Trust property).

Above right: **8.10** Lower Lincolnshire Limestone in typical small blocks in mortar, in Laxton (*c.* 1806).

Below: **8.8** The Manor House in Harringworth (late seventeenth century), built of Lower Lincolnshire Limestone, with quoins and dressings of Upper Lincolnshire Limestone, and Collyweston stone-slate roof.

ironstone is seen in several villages; Rushton has buildings of both. Along the edge of the Welland escarpment, limestone villages such as Easton-on-the-Hill and Gretton also display random blocks or stripes of the local ironstone, and Harringworth church has ironstone dressings – in an otherwise mostly Lincolnshire Limestone village (8.7, 8.8).

The limestone makes very neat rubblestone in small blocks (8.9), some only a few centimetres thick, some more square, and in many cottages well

8.11 The estate village of Laxton was built about 1806, of local Lower Lincolnshire Limestone, by George Freke Evans of Laxton Hall, and designed by Humphry Repton. Stafford Knot House was built as an inn, but is now a private house. It was originally thatched, but re-roofed with Collyweston stone slates between 1900 and 1910.

8.12 Typical masonry in Duddington: Lower Lincolnshire Limestone as coursed rubblestone, with dressings of Upper Lincolnshire Limestone oolite from Ketton.

shaped, making quite a smooth surface with mortar (see **8.10**, **8.11**). Quoins and window openings, however, are often made from one of the Upper Lincolnshire Limestone sources; in most places they are Weldon Stone, at Duddington they are of oolite from Ketton across the Welland valley (**8.12**).

When inspected closely, Lower Lincolnshire Limestone is usually seen as rather fine-grained sandy limestone, but it can otherwise have sparse or unevenly dispersed ooliths, sometimes, as at Laxton, with ooliths of a mixed range of sizes. The ooliths however are not prominent, and tend to blend with the fine-grained matrix of the rock.

CHAPTER NINE

Lincolnshire Limestone (2): the Famous Freestones

The Upper Lincolnshire Limestone in Northamptonshire is very localised (as explained in Chapter 2), confined to areas where channels cut down into the underlying beds, and patchy remnants here and there on the Lower Lincolnshire Limestone (see **8.1**). There would have been a greater thickness of it, but most of it was removed during the interval of erosion that followed soon after its deposition. The preservation of the buried channels and basins is fortunate, because it is from this Upper division that the best building stones have been obtained. The channelling of the sea-floor by scouring currents can explain some of the special features of these rocks: they are composed of ooliths and calcareous fossil debris in varying proportions, largely winnowed free of fine mud, giving rise to pure oolite or shelly oolite, some rocks remaining porous (indurated by the minimum of cement), but others having the intergranular space later cemented by sparry calcite. The existence of disparate channels during the build-up of the deposit further explains some of the differences recognised from one locality to another.

BARNACK

Barnack lies in the former Soke of Peterborough which was, until 1888, part of Northamptonshire, on an outcrop of Upper Lincolnshire Limestone extending some 2.5 kilometres towards Wittering (BGS map 157). The distinctive stone is known to have been quarried and transported by the Romans, (as recorded by Williams, and by Hill and others), and examples can be seen in the Museum of London; but the vast extent of the famous 'hills and holes' [TF076047] (a National Nature Reserve), together with documentary records and existing stonework, bear witness to a major industry in the medieval period, beginning with the Saxons. Morton in 1712 quoted an account of the early building of the Abbey of Medeshamstede (Peterborough) in 664 with Barnack Stone, noting: 'Such was the immense Weight and Bigness of some of the Stones that were laid in the Foundation of that Monastery....that eight Pair of Oxen cou'd scarce draw one of them'. Furthermore, 'The Abbeys of Ramsey, St. Edmundsbury, and all the Structures of ancient Magnificence all thereabouts, and to a great Distance, were built of Barneck Freestone'. The Abbey of Peterborough controlled the Barnack quarries, and allowed stone to be transported by waterways to the fenland abbeys (Ramsey, for example having to supply in return, 4000 eels during Lent). Successive early monastery buildings at Peterborough were eventually replaced after the disastrous fire of 1116 by the present magnificent Norman nave, the west front was added in the thirteenth century, and the porch in the fourteenth, in robust Barnack Rag. Since the apparent exhaustion of the Barnack quarries by the fifteenth century, repairs have been carried out in other types of Lincolnshire Limestone.

Two kinds of limestone were quarried: a cemented oolite with little or no shell, which was used for certain carved work (the Saxon eagle and the later effigy of Sir John de Verdun at Brixworth are said to be Barnack oolite); and the more distinctive Barnack Rag, packed with assorted shelly fossil debris including small gastropods and echinoderm spines, and ooliths which tend to weather out of the very strong spar cement, leaving holes (**9.1a,b**). It is cross-bedded, and the stone often has a ribbed appearance, the layers having different proportions of ooliths, but it is robust and has proved extremely durable.

The parish church at Barnack, built of Barnack Rag, has an early eleventh-century tower with characteristic Saxon quoins and long pilasters. The well-known Saxon tower of Earls Barton Church has similar features, as well as balustered openings, which are constructed from identical Barnack-type stone (see **9.2**). Pilaster blocks, some 1.5 metres long, and diagonals are not merely surface decoration, but

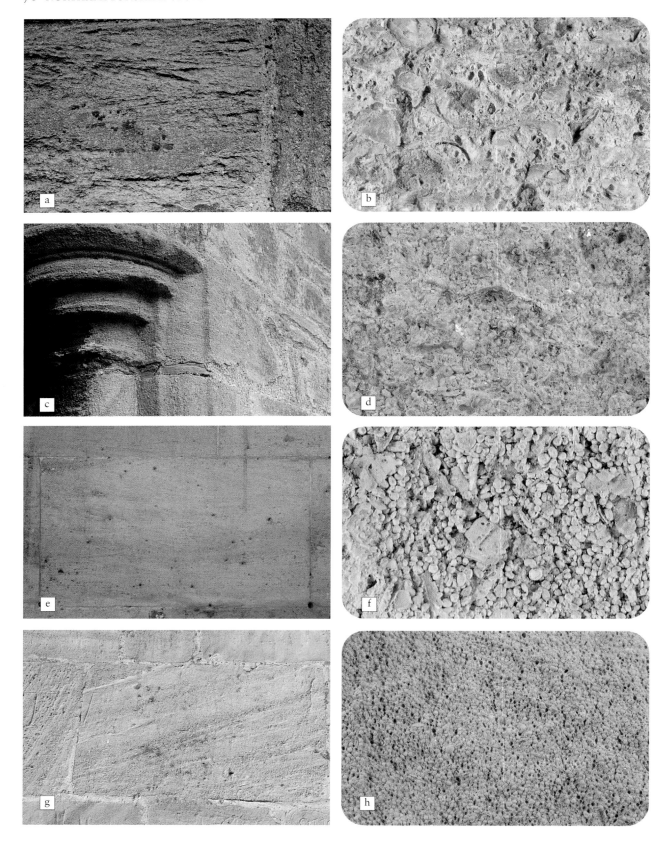

Opposite page: **9.1a-i** Varieties of Lincolnshire Limestone: on the left, stone as seen in a wall, on the right, stone as viewed magnified with a hand-lens.

Barnack Rag: a cross-bedded, banded limestone (Barnack Church): **b** sparry, shelly, oolitic limestone with varied fossil fragments; ooliths as holes in spar cement.

Stanion Stone: c cross-bedded, buff limestone (Stanion Church): **d** sparry, shelly, oolitic limestone. (LEIUG119523).

Weldon Stone: e cross-bedded, grey limestone with shell (Council Offices, Wellingborough). **f** Oolitic limestone, with shell fragments, porous; ooliths <1mm, with minimal cement.

King's Cliffe Stone: g cross-bedded, golden limestone (West Street, King's Cliffe). **h** Oolite, with shell fragments and bands of uneven-sized ooliths; porous. (LEIUG119517).

(*Above*): **i Ketton Oolite**, porous, ooliths 0.5-1mm, with minimal cement.

Above & below: **9.2** The Saxon tower of All Saints' Church, Earls Barton, with long blocks of Barnack Rag. Within the Barnack Rag framework the walling consists of local Wellingborough Limestone from the Rutland Formation (as seen in 1992, when the rendering was replaced).

are structural components within the walls, with local limestone from the Rutland Formation (see Chapter 10) packed between them (now invisible behind the replaced rendering). This large quantity of stone was perhaps dragged on sledges by oxen from Barnack down the old Roman route of Ermine Street 8 kilometres to the River Nene at Wansford, to begin a slow journey by shallow barge, more than 50 kilometres upstream. Saxon pieces of Barnack Rag at St. Peter's, Northampton, suggest that stone was also taken further up the river. Grave covers and stone coffins, requiring large blocks, were commonly made from Barnack-type stone, and have been found widely dispersed in archaeological contexts, not so long ago at Raunds.

Barnack Rag was primarily employed for medieval religious buildings; it was a stone suitable for sturdy masonry, but also able to be carved, the finest examples being without doubt the remarkable Norman doorways in Ely Cathedral. The Norman church of St. Kyneburgha at Castor, just 8 kilometres from Barnack, is built of Barnack Rag, and it is also recognisable in medieval architectural features (doorways, piers) in many churches in northern Northamptonshire which are otherwise built of their local stone – at Easton-on-the-Hill, Duddington, and Nassington (where it is also seen in the Prebendal Manor), all within 10 kilometres; Warmington, Oundle, and Barnwell are more distant, but alongside the Nene, which was surely a route for stone transport. (The stone of the Norman west door of Cotterstock church is similar to Barnack, but more buff coloured, possibly coming from Wansford.) Doorcases and fonts reached Great Addington and Harpole (9.3), though it is not certain where they were made; movement of individual items was not dependent on river transport, and they are widespread (the font at West Haddon, the font and tympanum at Pitsford, for example).

The last quarrying at Barnack recorded by Peterborough Abbey was in 1454. The south chapel of Barnack church (1500) is quoted by Purcell as

9.3 The twelfth-century Norman font in Harpole Church; shelly, sparry, oolitic Barnack Rag. It has been suggested that it was carved in a Northampton workshop.

being 'in a different stone', but this only refers to an internal arch of Ketton oolite; the external masonry is on inspection not unlike Barnack. It may have come from the adjacent Walcot land, whose owner built the chapel. Quarrying was probably discontinued in any case after the Dissolution in the 1530s, but that in itself released a new source of Barnack Stone from dismantled ecclesiastical buildings; several consignments in Stamford for example, as described by Dr Eric Till in 1997, went to build Burghley House in 1555, and Purcell cites the re-use of fenland abbey stone at Cambridge.

STANION

Morton in 1712 saw at Stanion the 'Marks of great Age in its spacious Hollows: and in the Stone digg'd out of it, whereof the Churches thereabouts are built'. The stone can be seen in Stanion parish church, the Perpendicular tower built of large ashlar blocks, mostly a spar-cemented oolite with fossil debris, similar to Barnack Rag but not quite as coarse-textured (9.1c, d); most noticeable is the rich colour, the ooliths being often yellowish, the colour contrasting with the pale Weldon-type limestone used against the windows.

Brigstock, just 3 kilometres along the same valley, lies mostly on Blisworth Limestone, but with Lincolnshire Limestone immediately to the west, and the village has a mixture of both. The famous Saxon church is an assortment of limestone rubble, but tawny Stanion Stone is recognisable in aisle windows (fifteenth century), upper tower and spire (fourteenth-century), and apparently also the Norman doorcase and arcade piers; but the huge (1.3 metre) quoins of the late Saxon tower, and the tower arch, are different. These are a coarsely oolitic pale cream Lincolnshire Limestone containing a little fossil debris but lacking the sparry matrix of the usual Stanion Stone, so the source is uncertain.

Stanion Stone is recognised partly by its colour; in the tower of Geddington Church, for example, it is noticeably darker against the Weldon Stone. A little is similarly identifiable in the famous Eleanor Cross nearby (see 9.5). It is also seen in the great church at Fotheringhay (9.4). In 1434 Richard, Duke of York engaged a freemason, William Horwood, to build the 'new body of a kirk' but the detailed specification (quoted by Salzman) did not identify the materials that would be provided. Three types of Upper

Lincolnshire Limestone appear to have been used: the masonry and certain buttresses of golden, slightly ribbed spar-cemented Stanion-like stone, the windows in a pale grey Weldon-type oolite, and golden oolite from King's Cliffe elsewhere, including the tower parapet. Sandy Lower Lincolnshire Limestone was used in rubblestone blocking when the east end was later dismantled. Lowick Curch tower, crowned with an octagon, seems also to be of Stanion Stone, the quatrefoil frieze of Weldon oolite.

The attractive small town of Higham Ferrers is mostly built of the local Blisworth Limestone (see Chapter 11), but several other varieties of stone can be seen. The Bede House built by Archbishop Chichele in 1428 is conspicuous with its stripes of ironstone (see **11.17**); the dressings are in Stanion-type limestone here, and also in the College he founded nearby in 1431.

The tower of St. Giles, Desborough, is a strongly cemented cream-coloured shelly oolite, probably Stanion Stone; it was built in 1529, onto the earlier church of purplish Northampton Sand ironstone from local Desborough quarries. The fine lofty steeple of Kettering Church is certainly Lincolnshire Limestone (the spire, according to Steane, rebuilt in Weldon Stone); the tower is said to be Barnack Rag,

9.4 The glorious lantern-towered Church of St. Mary and All Saints, Fotheringhay, built in 1434 by Richard, Duke of York. The older chancel and collegiate buildings were removed in 1573. The stonework comprises varieties of Lincolnshire Limestone: Stanion-type limestone (slightly darker), with pale oolite, similar to Weldon limestone, at the windows. (King's Cliffe Stone has been recognised from the parapet and as blocks in the north porch; rubblestone includes Lower Lincolnshire Limestone).

but it is somewhat like Stanion Stone. The overland route from Barnack would have been at least 35 kilometres, roughly along our A43, passing the Stanion quarries which were only 9 kilometres from Kettering.

Several stone pits once existed around Stanion, but an extensive area of old quarries is shown on a map of 1730 (NRO) on the north side of the road to Brigstock, on land belonging to the Brudenells of Deene [*c.*SP919869]. Stone was dug from here about 1725 for a farmhouse in Deene Park. In 1977 it was still possible to find the overgrown quarries known as 'Lord Cardigan's Pits', now occupied by the by-pass. Two to three metres of shelly, sparry oolitic limestone were exposed, underlain by softer limestone.

WELDON

The pale limestone from Weldon near Corby is Northamptonshire's best-known building stone; it was a freestone, suitable for both ashlar and carving – and remarkably resistant to the ravages of the English weather. It may have been known to the Romans, though its identification in the Romano-British London Wall is uncertain; and the historian John Bridges recorded that according to tradition it was used for the old St. Paul's Cathedral built in 1037. From Tudor times, Purcell reports, a great deal of Weldon Stone was transported via the River Nene and waterways eastward to Cambridge, from a loading point at Gunwade near Castor. In King's College Chapel the masonry changes to Weldon above the earlier use of paler Magnesian Limestone, but inside, the soaring fan-vaulting completed in 1515 is Weldon Stone at its most glorious.

In Northamptonshire the earliest records of Weldon Stone are for building the massive gatehouse at Rockingham Castle in 1275, one mason Edward Geoffrey receiving payment of 8s. 2d. for cutting 700 freestones, the cost of carting amounting to 3s 6d. There can then be no greater contrast in style than the elegantly wrought monumental Cross at Geddington, one of those erected by Edward I to mark the funeral route of Queen Eleanor's body in 1290 (**9.5**). The stone was popular throughout northern Northamptonshire for Gothic window tracery and for many of the spires, and it was selected for prestigious Perpendicular church towers such as Titchmarsh. In Higham Ferrers churchyard

9.5 The Eleanor Cross at Geddington, erected about 1295. The delicate carving in Weldon Stone remains remarkably sharp. The darker layers are Stanion Stone, the statues another stone.

stands the supremely elegant school, built like a chapel entirely in Weldon Stone by Archbishop Chichele in 1422. The church at Whiston, close to the Nene, was built mainly of Weldon Stone in 1534; but on the whole this far south, dressed stone was more often obtained from closer sources in the Northampton Sand.

Weldon Stone is familiar, and easily recognised. It is pale cream, weathering light grey, with shell fragments of small oysters, often layered in diagonal cross-bedding (see **9.1e**). When magnified with a

9.6 One of the most interesting buildings in the county, the Elizabethan Kirby Hall was built (1570-1575) for Sir Humphrey Stafford, mainly using Weldon Stone. Later architectural modifications were made by Sir Christopher Hatton. The splendid courtyard inside is mainly Weldon freestone (but banded, golden King's Cliffe Stone can be recognised in the loggia and elsewhere). Externally *(below right)*, the masonry is local Lower Lincolnshire Limestone rubblestone, with quoins of Weldon Stone. (English Heritage property).

hand-lens it is seen to be composed mostly of ooliths, spherical grains about a millimetre in size, with some dispersed curved shell pieces 3 to 5 millimetres, and characteristically no visible cement. The rock is essentially porous (9.1f). It is apparently this porosity that enables the stone to resist rain and frost action. Weldon Stone has, for instance, always been favoured for chimney stacks.

Northamptonshire is known for its grand country houses, created by the wealthy landowners of the sixteenth, seventeenth, and eighteenth centuries, and surrounded by their gardens, parkland, and grazing sheep. Deene Park is still owned by the Brudenell family, descendants of Sir Robert who bought it in 1514. Later in the sixteenth century a large Elizabethan mansion was built onto the older house around a courtyard. The grey ashlar came from the Weldon quarries just 3 kilometres to the south, but the eastern stair turret as well as the courtyard porch, of more yellow stone, may distinguish the

work of King's Cliffe masons; seventeenth-century additions are of recognisable Ketton Stone.

Just two kilometres away is Kirby Hall; though now a ruin, its historic architectural elements are largely intact (9.6). The manor of Kirby changed hands at the Dissolution, and was eventually acquired by Sir Humphrey Stafford of Blatherwycke who began building this house in 1570. Carved stonework around the courtyard contains his initials and emblems, including the 'Stafford knot'. The spacious courtyard is faced in ashlar, mainly Weldon Stone; but in charge of the work was Thomas Thorpe, one of a family of stonemasons of King's

Cliffe, and some of the dressed stone, including the entrance loggia, has the pronounced banding and yellow hue of King's Cliffe Stone (described below). The main exterior masonry consists of small rubblestone blocks of local Lower Lincolnshire Limestone, with Weldon Stone ashlar for quoins, gables and chimneys. After Sir Humphrey's death in 1575 it was bought by Sir Christopher Hatton of Holdenby (Queen Elizabeth's Lord Chancellor), passing in 1597 to his cousin, another Sir Christopher Hatton, who entertained the court of King James here on several occasions, then to his son (of the same name), whose major alterations and embellishments in Weldon Stone between 1638 and 1640 are attributed to Inigo Jones. Kirby's gardens

9.7 'Consider that I laboured not for myself only'. The extraordinary Triangular Lodge at Rushton was built by the fervent Catholic, Sir Thomas Tresham, between 1594 and 1597. The masonry is local Lincolnshire Limestone and ironstone rubblestone, the dressed stone is Weldon Stone. Every intriguing detail is based on the number three, a pointed celebration of the Trinity [SP830831]. (English Heritage property.)

9.8 Tresham's Lyveden 'New Bield' was unfinished when he died in 1605. This evocative cruciform building combines Weldon Stone dressings with ashlar of local Blisworth Limestone near Oundle [SP983853]. (National Trust property.)

were renowned. However from the late eighteenth century it went into terminal decline and ruin. It is now well cared for by English Heritage.

One of the famous families in Northamptonshire were the Treshams, many of whom were important national figures. Sir William Tresham (originally at Sywell) bought the estate at Rushton near Kettering in 1437; his grandson John built Rushton Hall, of Weldon Stone ashlar, early in the sixteenth century; there are later gables around the courtyard with dates of 1595, 1627 and 1630 (the last two being after the Hall had been sold to Sir William Cokayne in 1619). John's great-grandson, Sir Thomas Tresham, born in 1534, was responsible for several interesting and unusual buildings. In 1578 he began the Market House at Rothwell, employing a well-known freemason, William Grumbold of Weldon, with stone given by Sir Christopher Hatton who now owned the Weldon quarries. It incorporates ninety heraldic emblems of Northamptonshire families, its purpose being the 'perpetual honour of his friends', a celebration of friendship.

In 1580 Tresham became a fervent Catholic, and he was to spend many years imprisoned on account of his recusancy. His extraordinary Triangular

9.9 Lamport Hall, north of Brixworth. The central 5-bay house was built for Sir Justinian Isham, to John Webb's design, c.1655, the rusticated ashlar in Weldon Stone. The addition to the left was by Francis Smith of Warwick in 1732, and that on the right by his son William in 1741, in matching stone.

Lodge, finished in 1597, was devised ingeniously on the theme of the Trinity, having three sides, each with three gables, decorated with trefoils, inscriptions and many details to absorb the mind and spirit of the beholder (9.7). It was built of coursed rubblestone from Tresham's own quarries, with dark bands of local calcareous ironstone between layers of varied Lincolnshire Limestone from Pipewell and Pilton; with Weldon Stone dressings. It was a Warrener's lodge, built for his rabbit-catcher.

The most evocative of his buildings stands alone in a field near Oundle, a hollow shell of perfect grey masonry in the shape of a cross, its windows vacant, and without a roof; this is Lyveden New Bield (9.8). Two types of stone here are easily recognisable, local Oundle Stone ashlar (Blisworth Limestone) looking rather darker than the Weldon Stone. It was designed by Robert Stickells, Clerk of Works to the Crown, the stonework carried out by Tresham's masons, the

9.10 The Talbot Inn at Oundle, built of Weldon Stone by William Whitwell in 1626.

family of Thomas Tyrell. Sir Thomas Tresham died in 1605, (two months before his son Francis was implicated in the Gunpowder Plot) and the building was never finished.

Boughton House, near Kettering, is the Northamptonshire residence of the Duke of Buccleuch and Queensberry. Here is a very large complex incorporating a Tudor great hall, successive buildings having been added to an original fifteenth-century monastery (using much Weldon Stone and one and a quarter acres of Collyweston Slate). The impressive north front was the latest phase of building, in the French style, by Ralph, 1st Duke of Montagu, in the 1690s.

In 1560 John Isham, one of a family of merchants, bought the ancient manor of Lamport. The present Lamport Hall was begun by his descendant, Sir Justinian, in 1655, with a five-bay mansion designed by John Webb, the son-in-law of Inigo Jones. A later

Justinian in 1732 added the library wing to the north, designed by Francis Smith of Warwick, continuing Webb's façade in matching Weldon Stone; the corresponding wing to the south was added in 1740 by Sir Edmund Isham who had succeeded his brother in 1737, the architect now being Francis Smith's son, William. Further additions were made in the nineteenth century without affecting the symmetry of the fine south front (see **9.9** on previous page).

These historic houses are all in the northern half of the county, reflecting the cost and effort of transporting the stone, mainly by ox-cart, from the Weldon quarries; transport by water on the River Nene was used where possible, particularly for projects beyond the county (presumably for the Banqueting House in London). By the late nineteenth century, rail transport enabled the stone to reach all parts of the country. Exceptionally, Horton House south-east of Northampton built in the eighteenth century, was faced in Weldon Stone, a distance of more than 50 kilometres from Weldon by whatever route. The house was demolished in the 1930s, but The Menagerie, a folly built of Weldon Stone and the local Blisworth Limestone, still remains.

There is a good deal of Lincolnshire Limestone in Northamptonshire's most charming stone town, Oundle, along with the local Blisworth Limestone (see Chapter 11). Weldon Stone can be seen in the parish church (which is a mixture of limestones) in the great west door, and the two-storey porch built by the Wyatts in 1485; as quoins in Paine's Almshouses in West Street (see **1.2**), with the decorative Elizabethan entrance gate (brought, it is said from Kirby Hall). The Talbot Inn was rebuilt in Weldon Stone in 1626 by William Whitwell (**9.10**); the White Lion (1641) is also Weldon ashlar, as is the front of Bramston House (1700), the pilasters being of Ketton oolite. Oundle School memorial chapel is a fine building of 1922-3, mainly in Weldon Stone, by A.C. Blomfield.

There are one or two Weldon-type buildings in most of the main towns: in Kettering, the Alfred East Art Gallery (by J.A. Gotch, 1913); in Wellingborough, Swanspool House built in 1778, now the main Council Offices; in Northampton, the Public Library in Abington Street (by Herbert Norman, 1910) looks similar to Weldon, also the handsome carved oolite of the Inglis building in Fish Street, (but the Victorian Guildhall, of many stones, has streaky Ancaster-type Lincolnshire Limestone).

The village of Weldon lies just east of Corby, sheltered behind a by-pass but surrounded by the desolate remains of Corby's ironstone-working industry, much of it now a vast lorry park. Within the village old hills and hollows, grassy or wooded, and cottages of rubble and ashlar, bear witness to its past. The most interesting building is Haunt Hill House, adorned with the initials 'HF' and the Masons' Company Arms, which was built in 1643 by Humphrey Frisby (**9.11**); a stonemason, he was the son of a King's Cliffe stonemason, and married to Elizabeth Grumbold of the Weldon stonemason family, as well as being connected by marriage to the Thorpes of King's Cliffe. Weldon's church was said by Morton to be built of Stanion Stone but there is

9.11 Haunt Hill House in Weldon village (1636-1643) was built by master mason Humphrey Frisby, and bears the Arms of the Masons' Company in the gable. Collyweston roof.

9.12 A Weldon quarry in 1892: the scene at the loading jenny, with quarry horse and trolley, and the traction engine. (Photograph: Northampton Museums.)

9.13 Weldon quarry in 1974. The freestone was split by drilling a line of holes, inserting two metal 'feathers' into each, and hammering plugs between the 'feathers'.

none there now; the tower which was rebuilt in the eighteenth century is all grey Weldon ashlar and the rather golden rubble of the south aisle is Weldon-type porous oolite, probably from near-surface local material.

The Upper Lincolnshire Limestone cut a deep channel at Weldon, and locally rests directly on the Northampton Sand ironstone (see 2.11). Taylor in 1963 recorded 7 metres of limestone, with oolitic freestone at two levels. A hard spar-cemented shelly limestone, known as Weldon Rag, occurs impersistently within the oolites. Though not much used for building it was at one time prized locally as 'Weldon Marble'; Morton in 1712 cited a monument in the church at Deene (see 1.7), and a chimney-piece in John Bridges' house which he much admired.

The history of the Weldon quarries has been described in detail by Jeffery Best and others of Nene College, Northampton. Traditionally quarries were worked as small pits, with picks and shovels, levers to prise up the blocks from the bed, and saws to cut the stone; spoil was heaped around, giving rise to the familiar 'hills and holes'. Quarries became larger in the nineteenth century with improved machinery and tools, and in 1880 the coming of the railway to within 3 kilometres of Weldon greatly increased business. Large two-handled saws were used to make vertical cuts from the top of the bed, and blocks prised from below, to be removed by a winch on a timber beam. The blocks were moved on horse-drawn sledges, or wagons on a tramway, to the loading jenny (9.12). From there they went by 'road engine' to the railway station or local customers. A quarry was as much as 10 metres deep, and the best stone at the lower level. There were also extensive underground workings in the lower freestone, pillars of stone being left to support the roof.

The quarries, long owned by the Finch-Hattons (Earls of Winchelsea and Nottingham), were sold to the British Steel Corporation in 1962, who leased the main quarry to Stamford Freestone Ltd. from 1966. The stone was now cut by the technique known as 'plug and feathers'. First a series of drill-holes were made vertically through the bed. Into the holes were placed pairs of flared metal 'feathers', then metal plugs were positioned and hammered in, causing the rock to split along the line of holes (9.13). Work continued here until 1977, when ironstone quarrying extended into the area. Another old freestone quarry was reopened, by Weldon Stone Enterprises Ltd., less than a kilometre to the south, but there was insufficient good stone to continue quarrying after 1985, and to everyone's regret, Weldon Stone ceased to be obtainable here; a very small stock remains for restoration work. However, the skills of the stonemason have a strong base in Northamptonshire, and this firm has expanded to work stone from other British sources, employing 17 men, with modern computer-controlled sawing machines alongside specialist banker-masons (see 1.11, 1.12).

KING'S CLIFFE

A stone very similar to Weldon, a porous oolite with a little shell, was quarried at King's Cliffe. To see it in an unambiguous context one must visit King's Cliffe itself – and allow plenty of time to explore one of the most interesting stone villages in the county (**9.14**). The first impression is of a warm golden limestone, and on the whole this colour distinguishes it from the pale grey Weldon limestone. Closer scrutiny shows that King's Cliffe oolite commonly has a banded texture, cross-bedded, with thin layers of fine-grained ooliths and coarser ooliths with shell (**9.1g, h**). In the old 'hills and holes' of the former Cliffe Park (**9.15**) are two other types of limestone: a fairly pale, fine-grained (0.2 to 0.3 millimetre) porous oolite, and a very coarsely shelly, spar-cemented rock containing urchin spines, spiral shells of gastropods, and small lumps of limestone.

'Cliffe' refers to the steep slope of outcropping limestone alongside the Willow Brook, and the Park of 1,160 acres on the north side of the brook was Crown property until 1517 when it was leased, and later bought, by the Cecils of Burghley. The late Dr Eric Till of Stamford studied records for the building

9.14 An early seventeenth-century cottage, gabled and mullioned, in golden King's Cliffe Stone.

9.15 Old quarries in the former Cliffe Park, King's Cliffe; stone from here was used for Burghley House from 1556 [TL015972].

9.16 'Books of piety are here lent to any persons of this or ye neighbouring towns'. The school established by William Law in King's Cliffe housed his library of theological books from 1752. It is built of King's Cliffe Stone.

of Burghley House from 1555 to 1587. During 1556, 93 tons of stone were 'raysed' from 'Clyffe Park Quarrye' at the rate of 2s. a ton, and men from several villages near Stamford were provided with bread, ale, butter, eggs and meat as payment for carting it.

Purcell describes the use of King's Cliffe Stone at Cambridge, beginning with some supplied for King's College Chapel in 1460. Dressed stone, sometimes with Weldon, was used at Trinity College from 1518, including the fountain in Great Court in 1601.

There were two well-known families of master masons at King's Cliffe, the Frisbys and the Thorpes. A monument to three generations of Thorpe masons (all Thomas) in the church is dated 1623. The third Thomas was the master mason who built Kirby Hall (see above); his son John 'layd ye first stone' there as a small boy. John Thorpe went on to become an architect and surveyor for Royal estates. Under the terms of his will the Thorpe Almshouses in Park Street were built in 1668, (they are local rubblestone, the quoins and windows apparently Ketton oolite).

King's Cliffe is chiefly known as the home of the devout scholar William Law, who was born at 12, West Street in 1616. He became tutor to Edward Gibbon (father of the historian), returning to King's Cliffe in 1740. He lived in Hall Farm opposite the church from 1744 until his death in 1761, and along Bridge Street are several of the school buildings and almshouses established by him and Mrs Elizabeth Hutcheson (9.16).

Stone for the village was quarried on both sides of the Willow Brook valley, some of it for rubblestone but a great deal of freestone for ashlar. The remains of small quarries can still be seen against the north side of the Apethorpe road, with houses now built in them. The end house (see 1.6) was built by a quarry owner in 1750, though grey in colour the stone has characteristic fine textural banding.

King's Cliffe Stone can be seen in other villages: Apethorpe church tower for example is built of large ashlar blocks, the rubblestone of the rest being mainly local Lower Lincolnshire Limestone. It is also recognisable, among other limestones, in the church at Fotheringhay; it is recorded (RCHME) that the bridge built there over the Nene in 1573 was of King's Cliffe Stone, and again when it was replaced in 1722.

KETTON

As Morton said, 'I cannot forbear to mention it, tho' belonging to another County'; Ketton Stone was much used in Northamptonshire towns (see 5.5), and is easily recognised. It is the purest oolite, with virtually no shell, and so little visible cement that one wonders how it holds together (9.1i). Masonry sometimes has a pinkish blush. The stone came into prominence rather later than the other varieties, in the seventeenth and eighteenth centuries, when it was eminently suitable for the pillars, pilasters, and fine ashlar of neoclassical buildings. Early eighteenth-century gravestones of Ketton oolite are decorative features of many a churchyard.

CHAPTER TEN

The Wellingborough Limestone, and Helmdon Stone

Within the sands and clays of the Rutland Formation is a limestone, usually sandy, often full of oysters, and generally only one to two metres thick; on current geological maps it is shown as 'Upper Estuarine Limestone' but it is now known as the 'Wellingborough Limestone' for most of the county, continuing as the Taynton Limestone in the south-west (10.1). In the cutting in Irchester Country Park it makes a ledge, separated from the Blisworth Limestone above by a few metres of clay (see 2.12). It has been used as rubblestone in villages near its outcrop, but may be accompanied by Blisworth Limestone, from which it is not easily distinguished. It forms a long narrow outcrop on the Great Doddington ridge, and in village walls the limestone is recognisable, packed with oysters and containing echinoderm spines, in a sparry matrix. The outcrop extends two kilometres west to Earls Barton, where it was quarried locally by the Saxon builders of the tower (see 9.2). Isham village has a great deal of shelly rubblestone in flat blocks (some in the church being almost a metre in length) and there is also a thin-bedded sandy limestone in local walls. At Pytchley this limestone was sufficiently fissile at outcrop, according to Morton in 1712, to be used as roofing slate.

To the west of Northampton patches of Blisworth Limestone occur on high ground each side of Dallington brook; on Hopping Hill it was quarried for limekiln burning. Below it a ledge of oyster-rich sparry Wellingborough Limestone protrudes among clays of the Rutland Formation that were once dug for brick-making. The limestone forms rubble walling in the churches at Dallington and Duston. Some of the houses in both these villages are banded, with courses of Northampton Sand sandstone and the oyster-rich limestone (10.2). Harpole village has a great deal of imported ginger Duston sandstone, but the tower of the church is different, being random rubble of flat blocks 3 to 10 centimetres

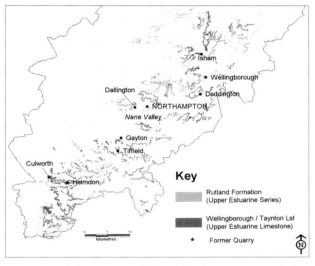

Key

    Rutland Formation (Upper Estuarine Series)

    Wellingborough / Taynton Lst (Upper Estuarine Limestone)

★   Former Quarry

10.1 Map of the outcrop of the Rutland Formation with the limestone (former Upper Estuarine Limestone) now known mostly as the Wellingborough Limestone - but in the south-west becoming the Taynton Limestone. (On this map 'Lower Estuarine Series' is included with the Rutland Formation).

    IPR/37-3C British Geological Survey. © NERC. All rights reserved.

10.2 Several houses in the village of Dallington (now part of Northampton) are banded, using local stone: brown sandstone from the Northampton Sand, and pale limestone, full of oysters (former Upper Estuarine Limestone, now known as the Wellingborough Limestone).

10.3 A delightful old house in Gayton village, built of Wellingborough Limestone (former 'Upper Estuarine Limestone'), available just south of the village, with quoins and mullioned windows of brown sandstone, from the Northampton Sand.

thick, some up to 30 centimetres long, of shelly, sparry limestone, which probably came from the Rutland Formation around Duston.

South of the River Nene, Rothersthorpe Church has a similar shelly, sparry limestone, containing a little ochreous material; it can look dark on account of the lichen it attracts, but bare or broken edges show it to be quite pale, unlike the iron-rich Marlstone Rock from the local outcrop that can be seen in some cottages. Bugbrooke church tower is banded, having courses of Northampton Sand and the shelly limestone. The nearest outcrops of the limestone lie about 3 kilometres to the south of these villages, on the crest above valleys leading to the Nene. Gayton is beautifully situated high above a valley, occupying an outcrop of Northampton Sand,

but a great deal of the village is built of the Wellingborough Limestone from the Rutland Formation that occurs just to the south (10.3). The limestone continues, emerging from a cover of boulder clay at Tiffield three kilometres further south. This is another village largely built of the oyster-shell limestone.

Towards the south-west of the county the limestone gets thicker, and outcrops widen near the villages of Helmdon and Weston, Sulgrave and Culworth, Greatworth and Farthinghoe. Here geologists refer to it as the Taynton Limestone Formation. A quarry near Culworth was known to Morton and Bridges in the early eighteenth century, supplying smooth white stone for flooring in the big houses of the district - such as Canons Ashby in 1710 (see 3.8) and later, Edgcote House in 1752. The flagstones often have echinoid spines.

Culworth village, lying on Northampton Sand, has some sandstone in buildings, but otherwise the villages are mainly built of the local limestone. At

10.4 Easton Neston House, built for Lord Leominster in 1702, by Nicholas Hawksmoor, following Wren's earlier wings in brick. It is faced with Helmdon Stone ashlar, the stone also used for the giant Corinthian pilasters. Seen close up (*right*), the limestone characteristically has streaks of fine powdery cream micrite.

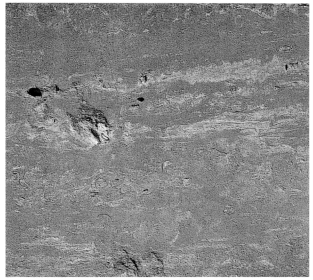

Farthinghoe the stone includes shelly, spar-cemented and also oolitic varieties, some laminated and flaggy, containing spines and shell fragments. But around Helmdon some of the limestone is in thick beds, which in the past were quarried for freestone.

HELMDON

At the beginning of the eighteenth century Morton noted the four 'Quarries at this time of chiefest Note with us, are those of Halston, Helmdon, Welden and Rance'; of Helmdon, the stone used to build Easton Neston House about 1700, he said that the 'fair white and durable Stone...is indeed the finest building Stone I have seen in England'. The House (10.4) was built for Sir William Fermor, MP for Northampton, who became Lord Leominster in 1692; two brick service wings were built first, by Christopher Wren, of which the north one remains, but the main House built by 1702 was the work of Wren's assistant, Nicholas Hawksmoor, who modified earlier plans, and created a masterpiece.

The limestone is geologically continuous with the more famous Taynton Limestone quarried in the Cotswold village of Taynton near Burford in Oxfordshire which (as Arkell described) went to many of the Oxford Colleges, to Windsor Castle, and to Blenheim Palace. The limestone is much thicker in the Cotswolds, and Taynton Stone is a strong, sparry oolitic limestone, with varying amount of shell. By southern Northamptonshire, a distance of 50 kilometres, the limestone of Helmdon is only 3 metres thick, and differs from the Cotswold stone. It is cross-bedded and sandy, with streaks of

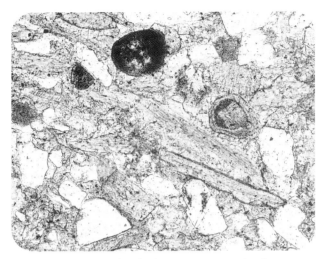

10.5 A thin section of Helmdon Stone seen under the microscope. It has been stained, to show oyster shells (pink, each shell composed of many thin layers), quartz sand grains (colourless), and a matrix of calcite grains (stained blue). The field is about 3 millimetres.

10.6 A rubblestone cottage in Helmdon village, with dressed stone doorcase.

cream-coloured fine powdery material (micrite), and contains a great deal of broken shell, many oysters, and urchin spines, the matrix finely granular rather than sparry (10.5).

Helmdon village buildings (10.6) are of rubblestone - some have stone architraves, for the village was home to families of stonemasons; E.G.Parry's historical account of Helmdon Stone mentions there being as many as 10 masons here about 1730. William Wigson, a mason from Eydon, bought property and land in Helmdon in 1688; two of his kinsmen, John Wigson (see page 56) and Joshua (both of whom witnessed the indenture), apparently went with him; the property is still known as Wigson's Farm. William died in Helmdon in 1727; Joshua, a freemason who worked on Blenheim Palace, died in 1740.

Helmdon quarries were on each side of the road to Weston (10.7). A little Helmdon Stone had been used for a mansion at Stowe in Buckinghamshire in the 1670s, but Parry reports that much more was transported there after 1710 for the eighteenth-century rebuilding by Sir Richard Cobham, and the many temples erected in the gardens. Easton Neston garden also has a temple of Helmdon Stone. At Delapré in Northampton a rusticated garden gateway of Helmdon Stone makes an architectural feature (see page 4).

Between 1705 and 1710 Blenheim Palace was being built for the Duke of Marlborough in Oxfordshire, mainly of stone from Taynton and Burford, but Parry points out that consignments were also obtained from suppliers in Helmdon (traditional stoneworking families with names such as Wigson, Bayliss and Stockley) and a total of £600 was paid for stone carted from Helmdon to Blenheim, some of it for floor paving. Later in the eighteenth century, paving stone was also supplied from Helmdon for Woburn Abbey. But while some stone was going to Blenheim, good Helmdon ashlar was used to build Brackley's Town Hall, erected by the Duke of Bridgewater in 1706 (10.8).

At Canons Ashby House the alterations carried out by Edward Dryden in 1708-10 included the insertion of a fashionable doorcase with broken pediment in Helmdon Stone on the south side. But here at Canons Ashby is evidence of Helmdon Stone's much longer history, reaching back to the thirteenth century. The church is a remnant of a former Augustinian priory, built of ferruginous

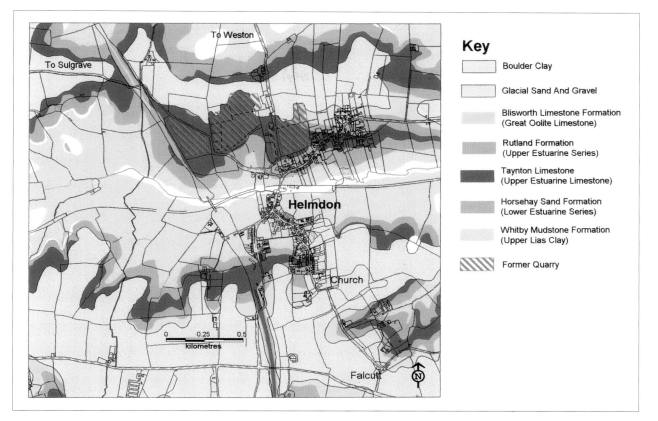

10.7 Geological map of Helmdon, showing the quarries in the Taynton Limestone (former Upper Estuarine Limestone). IPR/37-3C British Geological Survey. © NERC. All rights reserved. The quarries are as outlined by David Hall in Parry's 1987 account.

Right & below: 10.8 The Town Hall at Brackley was built in 1706 of Helmdon Stone ashlar. The limestone, seen on the bed, contains oyster shell fragments and urchin spines.

Above: **10.9** A stained glass window as early as 1313 in Helmdon Church is inscribed with the name of William Campiun and depicts a medieval stonemason at work.

Left & below: **10.10** The Eleanor Cross in London Road, Northampton (Hardingstone parish) [SP754583], was carved in Helmdon Stone (*c.*1294) by John de Bello, the four beautiful statues possibly of Barnack oolite, by William of Ireland. The cream Helmdon Stone is typically cross-bedded and streaked with finer powdery micrite. Parts of the monument have been restored with Ketton oolite.

sandstone from the local Northampton Sand, with blank arcading on the west front carved in Helmdon-type limestone.

Other churches in the south-west are also embellished with this creamy-grey limestone full of oyster shell, recognisable in the west portals of Middleton Cheney and Byfield. (The south porch of the former, however, has moulding of different, sparry oolitic Taynton-type stone, and limestone at Kings Sutton also includes sparry oolite from the Cotswolds.)

In Helmdon village the medieval church is, of course, built of the local stone - the external masonry, and the internal arcade piers; but unexpected testimony is to be found in a tiny stained-glass window of 1313 in the north aisle, inscribed with the name of William Campiun, and depicting a medieval stone-mason (10.9).

Northamptonshire is fortunate in having two of the three surviving crosses erected by Edward I in memory of his Queen, Eleanor of Castile. (There are thought to have been twelve, marking most of the nightly resting places of her coffin on its 159-mile journey from Harby in Nottinghamshire to Westminster, in 1291.) They are masterpieces in stone: the cross at Geddington (see 9.5) was carved in limestone from Weldon and Stanion; the one in Northampton, close to Delapré Abbey, was made of of Helmdon Stone (10.10). Alice Dryden quoted Executors' records of payment to John de Bello, 'cementarius', who carved the cross, and William of Ireland the 'imaginator' who made the statues, also to 'William of Bernak, mason, for the carriage of four images to the cross of Northampton', which suggests that the statues themselves may be of Barnack oolite.

(Stones mounted in the wall nearby, formerly under the statues, appear to be Weldon Stone.) The Northampton cross has been restored several times and, strangely, Ketton oolite rather than Helmdon Stone was used for repairs in the eighteenth century. However, the streaky, shelly Helmdon limestone can still be recognised. The reader may like to compare this stone with the limestone at St. Peter's Church in Northampton (see **frontispiece** and **11.11**). St. Peter's is discussed in Chapter 11, in weighing up the likely contenders from the wider field of the Blisworth Limestone; Helmdon is also among those that have been considered.

Helmdon seems to have seen active quarrying in two separate periods, the medieval mainly ecclesiastical work coming to an end apparently before the fifteenth century. Whereas Weldon Stone continued to be popular for secular buildings in the north of the county, there is at present little evidence that Helmdon Stone was used for anything more than village building until perhaps late in the seventeenth century. The stone was then found to suit the style of fine ashlar, pilasters, columns and carving.

Use of Helmdon Stone reached its second peak in eighteenth-century architecture. By the nineteenth century it was in decline, and though Parry records some stone going to Canons Ashby in 1829 he observes that four masons listed in the Census of 1841 have dwindled to just one by 1871. An area of grassy hollows is all that remains of Helmdon's extensive quarries; the sawmill [SP587443] occupies one of them, just a metre of upper limestones being visible in a cutting, with several very large oyster-limestone blocks nearby that were obtained from the site.

CHAPTER ELEVEN

The Blisworth Limestone Formation: with Freestones, 'Marbles', and Polychrome Stripes

This is the common limestone of southern and eastern Northamptonshire, extending from Aynho and Brackley in the south in a wide sweep north-eastwards to beyond Oundle. The outcrop however (11.1) looks somewhat sketchy, because much of it is covered by boulder clay, deposited over the landscape of Jurassic rocks in the Ice Age; and north of Thrapston there are other Jurassic rocks above the limestone. The rivers have now cut down through the glacial cover, exposing the underlying Jurassic rocks along the valleys, and it is mainly around the tributaries of the Tove in the south, and the Nene to the north-east, that the outcrop of the Blisworth Limestone produces its intricate contoured pattern. The expanses of boulder clay are only sparsely populated, but close to the limestone outcrop are more than 80 of the county's villages and one or two small towns. They mostly had their own stone-pits; though few exist today, the stone is there to see in the village walls, an accessible reflection of the geology below ground. Away from the main outcrop, Church Stowe with Upper Stowe, south-west of Northampton, stand out as limestone villages in mainly brown sandstone territory. Here is a narrow strip of Blisworth Limestone brought down between two geological faults. An old quarry still exists (it provided limestone for the local iron-works in the nineteenth century), alongside a quarry in brown Northampton Sand sandstone. The limestone (possibly also the sandstone) may have been taken from here to Bugbrooke, a village three kilometres away, on Lias, where there are several fine old farmhouses of Blisworth Limestone, with window dressings of brown Northampton Sand.

The limestone villages are mostly cream or grey rubblestone. Usually the limestone has varying amounts of shelly fragments, especially oysters, and cross-bedding is common (11.2a, b). But often some of the limestone is soft and powdery, and some

walling blocks, or parts of them, are poor-weathering (11.2c). Though it was formerly known as the 'Great Oolite Limestone', in Northamptonshire it is not always oolitic. Instead, many bits of shell may be coated by a layer of fine carbonate which makes them smooth and often oval-shaped (they are 'superficial ooliths', flatter than proper ooliths which are spherical and made of concentric layers). They can be seen with a lens (11.2d); the Raunds limestone has spar cement, while Cosgrove (11.2e) is more granular. Rubblestone is often a mixture of types, but there are regional variations. Oundle limestone (11.2f) is commonly shelly and sparry, while thinly bedded (laminated) limestone (11.2g) is a variety seen in southern Northamptonshire. Apart from local stone-pits providing rubblestone, there were in the past a few places with well-known quarries producing freestone.

COSGROVE

According to Morton (1712), 'The Quarries of most ancient Note in all the Southern Parts of the County are those at Cosgrave, which have been digg'd under Ground in the form of Caves or Vaults; large Stacks of the Quarry-Stone being left standing at due Distances to support the Roof'. The hills and hollows of former quarries can still be seen south-west of the village. In 1927, Beeby Thompson said he could remember two doorways giving access to the underground workings from the rock face on the south side of the road but they were no longer accessible; two metres of building stone however could still be seen below three or four metres of overlying limestone beds (this quarried area south of the road was finally infilled in the 1980s).

The quarries were ideally located on the valley slope just 700 metres from the River Great Ouse,

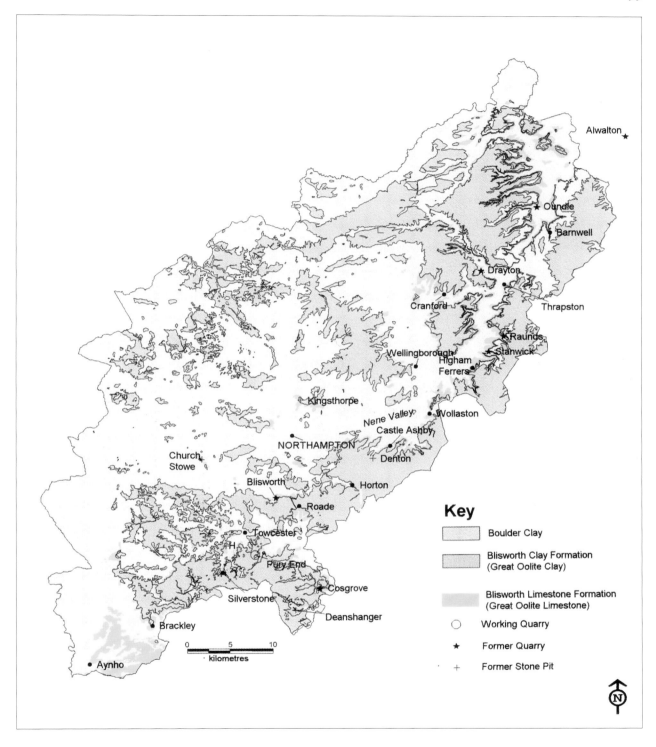

Key

�	Boulder Clay
	Blisworth Clay Formation (Great Oolite Clay)
	Blisworth Limestone Formation (Great Oolite Limestone)
○	Working Quarry
★	Former Quarry
+	Former Stone Pit

11.1 The outcrop of the Blisworth Limestone (Great Oolite Limestone) is shown in yellow. Note that it also underlies the outcrop of relatively thin Blisworth Clay, and in many places may be covered by Boulder Clay. (Geological maps of one inch or 1:50,000 show Jurassic boundaries below the Drift.)

Over 80 villages and towns near the outcrop in southern and eastern Northamptonshire are built of this limestone. H Handley Park (see text).

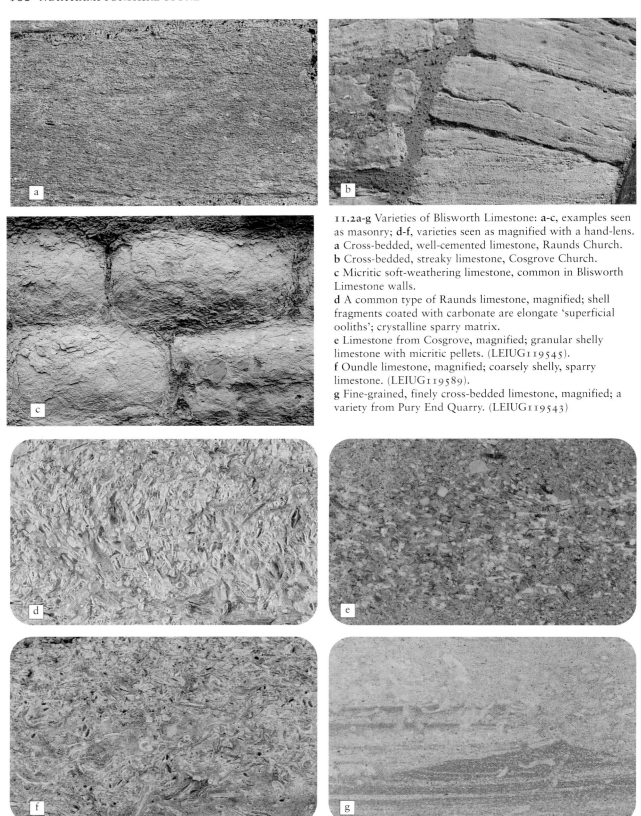

11.2a-g Varieties of Blisworth Limestone: **a-c**, examples seen as masonry; **d-f**, varieties seen as magnified with a hand-lens.
a Cross-bedded, well-cemented limestone, Raunds Church.
b Cross-bedded, streaky limestone, Cosgrove Church.
c Micritic soft-weathering limestone, common in Blisworth Limestone walls.
d A common type of Raunds limestone, magnified; shell fragments coated with carbonate are elongate 'superficial ooliths'; crystalline sparry matrix.
e Limestone from Cosgrove, magnified; granular shelly limestone with micritic pellets. (LEIUG119545).
f Oundle limestone, magnified; coarsely shelly, sparry limestone. (LEIUG119589).
g Fine-grained, finely cross-bedded limestone, magnified; a variety from Pury End Quarry. (LEIUG119543)

and equidistant from Watling Street. They may have been started by the Romans (a Roman building stood on the same slope, about a kilometre to the east), but they were probably at least medieval. Today old buildings in the village provide the only examples likely to be local Cosgrove stone. The church has some Norman work in the chancel and a fourteenth-century tower with mouldings (**11.2b**). The stone is cream-coloured and streaky, cross-bedded, with thin layers of coarser and finer texture, composed of granular shelly debris (oysters and other bivalves, occasional urchin spines) in a not very prominent matrix. Some stone encloses lumps of fine-grained white limestone; the rubblestone also includes a laminated variety.

Cosgrove Hall (**11.3**) was built in the early eighteenth century, perhaps (as noted by Pevsner) by master-mason John Lumley of Northampton. It is faced in cream ashlar, the blocks streaked by cross-bedding, composed of shelly material similar to the church masonry. The pilasters are of long flat slabs, the layering up-ended. Cosgrove Stone is known to have been used for quoins and doorcases at Winslow Hall in Buckinghamshire, a brick mansion built in 1702 and attributed to Christopher Wren. The Grand Union Canal runs through Cosgrove beneath an architecturally interesting bridge built in 1800,

11.3 Cosgrove Hall (early eighteenth century) may have been built by master mason John Lumley of Northampton (1654-1721), for barrister Henry Longueville. The masonry is ashlar of cross-bedded local Blisworth Limestone, Cosgrove Stone.

mostly in coarsely granular golden limestone with ribbed weathering, possibly a local variety.

Beeby Thompson, reporting on the county's limestone quarries in 1927, mentioned one for building-stone west of Deanshanger, and he cited Baker's reference to Handley Park as the source of stone for Towcester Church by the gift of Edward IV. The stone obtained, however, as one can see from the Church, was ironstone, and very little limestone. The only quarry working in the area now is at Pury End (see **1.9**).

BLISWORTH

Though Blisworth Limestone is naturally much in evidence, the village is particularly admired for its conspicuously striped cottages (see below). The church is the oldest building in the village, its thirteenth-century octagonal piers and doorway mouldings possibly providing evidence of early stone-working here. The external masonry includes streaky, cross-bedded cream limestone of granular shelly material, with the usual oysters, in a rather soft powdery matrix. Some is more sparry, with elongate, coated shell grains and some ooliths.

But little is known about Blisworth as a source of quarried stone before the nineteenth century. No quarry is mentioned by Morton or by Bridges. In 1799 the Grand Junction Canal was constructed as far as the difficult hill section between Blisworth and Stoke Bruerne, and it was not until 1805 that the Blisworth tunnel created a continous waterway. The Duke of Grafton opened up limestone quarries close to the tunnel in Blisworth in 1821, and the Blisworth Stoneworks building was constructed there about 1834 (11.4). A tramroad gave access to the canal. Samuel Sharp described the limestone quarry in 1870 as 'ancient and large', furnishing various limestones for building and lime-burning. The limestones (lying beneath more than 3 metres of boulder clay) included at the top 1.5 metres of 'Pendle' [not the Northampton Sand 'Pendle'], a hard limestone for rough building; a metre of sandy limestone ('Sandstone') - a 'good building stone' - between beds of marl (burnt for lime); and 0.75 metre near the bottom called 'The Blocks', a 'freely working oolitic limestone - sawn up and faced for flooring, window-sills and chimney-pieces'. By 1902 the limestone was being quarried mainly for flux for the ironstone furnaces at Hunsbury, the wagons tipping it into barges on the canal alongside the ironstone quarried close by. The limestone quarries then occupied about 24 acres, but quarrying finished in 1921 with the closing of the Hunsbury furnaces, and eventually they were filled in.

There was a stone-pit south-east of Denton beside the Northampton-Bedford road which was a local source of limestone, on the Castle Ashby estate. One reads conflicting statements about the stone of the great House at Castle Ashby (11.5). Pevsner says 'it is built of Weldon Stone' (as had Bridges who was, I think, referring to the parapet), while more recently Heward and Taylor say it is 'Ketton stone ashlar'.

11.4 Blisworth limestone quarries were opened close to the Grand Junction Canal in 1821, and the Blisworth Stoneworks built in 1834. Quarrying ceased in 1921 and it is now a farm building.

The House, built on the site of an earlier castle by the first Lord Compton in 1574, is in fact made of local Blisworth Limestone rubblestone, initially with quoins of ironstone, and windows of Weldon Stone. The elegant screen in the centre of the south front, added between 1625 and 1635, with ashlar facing on the garden side to the east, is recognisably Weldon Stone. The local limestone contains plenty of bits of oyster shell and a few ooliths, some is cross-bedded, and some fine-grained and powdery, occasionally containing large oysters; in some of the village walls there are shoals of brachiopods.

STANWICK AND RAUNDS

The Blisworth Limestone continues along the side of the Nene valley, with the hilltop villages of Grendon, Wollaston, and Irchester built of it. Higham Ferrers is one of the county's attractive small towns, with some historic buildings of quite good shelly limestone. There were local 'Town Pits', and also quarries to the north towards the village of Stanwick ('Stanwige' in Domesday Book of 1086). The prefix 'Stan' indicates a place long associated with stone. There were many quarries here, some of which were last used to supply flux for the iron furnaces, but their history is probably medieval, or earlier. A Roman villa, which was excavated in the 1980s in the valley below Stanwick, certainly used local

Above & below: **11.5** Castle Ashby House was built by Lord Compton *c.*1574, of local Blisworth Limestone rubblestone with ironstone quoins, and Weldon Stone for windows. The grand screen in the centre of the south front, probably by Inigo Jones, was added *c.*1630, along with alterations to the east elevation, in Weldon Stone. The masonry seen close up in the south-east corner is Blisworth Limestone, the quoins having been replaced by Weldon Stone during alterations to the east range and the chapel, in 1624.

limestone. Though most buildings are of the usual rubblestone, there is dressed stone in Stanwick Church and in the superb thirteenth-century church at Higham Ferrers, where the entrance mouldings, albeit now worn, are shaped in Blisworth Limestone. The tall crosses in Higham churchyard and market place are of beds of the limestone set on end. Higham Ferrers Castle, long demolished, would have required large quantities of local stone, but Stanwick is not specifically mentioned in Duchy of Lancaster accounts, which only begin with repairs in the fourteenth and fifteenth centuries; W.J.B. Kerr records some stone from Rushden, and 'sclatstone from Yerdele' (both from the Blisworth Limestone), 'John Evan of Yerdele [Yardley], sclatier', being paid £5. 3s. 4d. for making a stone roof. Weldon Stone and slates from Kirby (Lincolnshire Limestone) are also listed in the accounts.

The quarries at Stanwick and in neighbouring Raunds were known in the seventeenth and eighteenth centuries as sources of 'marble' (see below), and 'Rance' (Raunds) was particularly mentioned by Morton in 1712 as being one of four quarries 'of Chiefest Note' in the county at that time. But you have only to see the magnificent parish

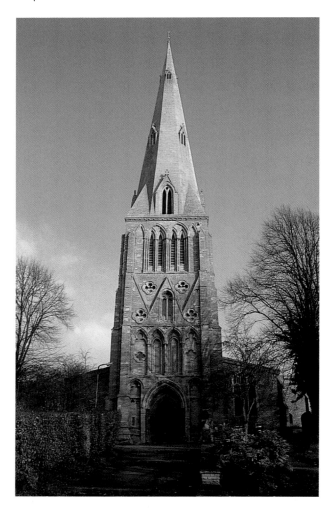

11.6 The very fine Church of St. Mary, Raunds, built of the local Blisworth Limestone, including *(above)* the elaborate Early-English mouldings of the west tower in Raunds freestone.

hillside west of the Raunds brook, where small quarries in Blisworth Limestone, and shallow pits dug in the overlying Cornbrash outcrop, provided rubblestone for a small stone church which was built in the late ninth or early tenth century, rebuilt in the eleventh or twelfth century, and then converted to secular use as part of a manor house by the thirteenth century. The Blisworth Limestone quarries were separate small pits, about 6 metres by 4 metres, and 1.7 to 2.1 metres deep; they were backfilled during the eleventh and twelfth centuries. On the east side of the Raunds valley, north of the parish church, seven more quarries were found, the largest

church at Raunds, one of the best examples of Blisworth Limestone in the county, to appreciate that here was building stone with a long history. The Early English tower (11.6) has ashlar buttresses and an elaborate west portal with arch-heads in Weldon stone, but most of the moulding here and at higher levels, is in local Raunds stone. It is cross-bedded in thin layers, shelly and containing some ooliths with many flatter, coated shell grains ('superficial ooliths') in a good spar cement (11.2d) - somewhat resembling the appearance of Barnack Rag but lacking Barnack's more varied fossils. It is cream rather than grey, and has some of the powdery character typical of the Blisworth Limestone.

Archaeological excavations in Raunds in the 1980s by Graham Cadman and others in the Northamptonshire County Council Archaeological Unit uncovered a remarkable series of quarries. Some as old as Late Saxon (850-1050 AD) were on the

11.7 Cranford is one of the many villages built of local Blisworth Limestone. Good-quality rubblestone houses often have quoins and windows of oolitic Weldon Stone (Lincolnshire Limestone).

11.8 Barnwell Castle was built in 1266, of local Blisworth Limestone.

being 8 metres by 6 metres and 2 metres deep, again backfilled in the eleventh or early twelfth centuries. Later medieval quarries (infilled between 1350 and 1550) were identified in the hillside outcrop of Thorpe End to the south; next to them a factory site on the corner has been enlarged in another former quarry, of unknown date.

Many villages of Blisworth Limestone lie on the fringe of high ground east of Kettering, and in the valleys leading to the Nene (11.7). Near Thrapston the limestone outcrop narrows under the cover of overlying Jurassic rocks. The Cornbrash forms a continuous higher outcrop along the east side of the Nene valley, but Blisworth Limestone was still accessible for villages such as Barnwell, from the thirteenth-century castle (11.8) to later cottages. Cornbrash limestone was apparently not much used for building, but a little yellowish flat rubblestone, composed of fine shelly brash, is recognisable in Barnwell walls, and more so in Polebrook parish church.

Lilford Hall, built between 1632 and 1635, and standing close to the Nene south of Barnwell, is an example of the high-quality Blisworth Limestone coming in towards Oundle. The grand front of the Hall (11.9) combines good Blisworth Limestone ashlar with gables, bay windows, and window-cases of paler Weldon oolite (see also 1.4). There used to be a quarry at Lilford Lodge Farm, with several beds of massive, cross-bedded, shelly, sparry limestone (see 1.10). Nearer Oundle, large blocks of limestone remain on the river bank from quarrying sometime in the past.

11.9 Lilford Hall, south of Oundle, built in 1635 for William Elmes, of local Blisworth Limestone ashlar (Oundle type), with gables and windows of paler Weldon Stone (Lincolnshire Limestone). (See also 1.4).

OUNDLE

This, the most attractive stone town in Northamptonshire, is built on a bluff of Blisworth Limestone in a sharp bend of the River Nene, the glorious spire of the parish church visible from afar. When Leland in 1530 observed that the town was 'all buildid of stone', much of it would have been local Blisworth Limestone, but the church also includes thirteenth-century arcades of a robust oolitic limestone that probably came from Barnack, the great west door to the fourteenth-century tower is made of Weldon oolite, and so is the ornate two-storey porch, erected in 1485. The following centuries saw the building of some handsome houses, for this was a prosperous town. Though many were embellished with Lincolnshire Limestone from Weldon and Ketton, the architecture of Oundle could rely upon the excellence of its own Blisworth Limestone, which provided not just rubblestone, but good ashlar. There were many quarries in the present area of the town, and more out on the road to Benefield where hills and hollows underlie the golf course. Until recently a quarry was worked at Churchfield, not far from the standing ruin of Lyveden New Bield which is largely local Oundle-type ashlar (see **9.8**).

Oundle Stone is full of shells, closely packed in a good sparry matrix (**11.2f**). A lot of it can be seen in the town, beginning with the row of seventeenth-century rubblestone cottages in Mill Road, and on to the old building known as Paine's Cottage or Almshouses in West Street (see **1.2**). The fine house known as Cobthorne (**11.10**) was built in 1656 by William Butler, one of Cromwell's officers, from Oundle ashlar with dressings of Weldon oolite. Another building of the same period is the bookshop, with walling of Blisworth Limestone, and an unusual colonnade made of oolitic Ketton stone. Across the Market Place is the house built about 1700 for a lawyer, Stephen Bramston; its ashlar façade is mainly Weldon, with giant Doric pilasters of Ketton oolite, but the walling to the side is well-dressed local limestone. In the south-west corner of the churchyard a small Georgian house built for the Laxton schoolmaster in 1763 is entirely faced in Oundle ashlar, and so is Oundle School's 'New' schoolhouse built next to it in 1799. Later school buildings introduced more Lincolnshire Limestone from Weldon and Ketton (see Chapter 9).

This brief account cannot look in more detail at the town, but an excellent 'Town Trail' introduces the history, the buildings and the people; and a hand-lens will help to sort out the limestones.

11.10 Garden-side view of Cobthorne in Oundle, a fine ashlar house of shelly Oundle Stone built by Major-General William Butler, one of Cromwell's officers, in 1656. The quoins and windows are of paler Weldon Stone.

LIMESTONE IN NORMAN NORTHAMPTON

The town's perhaps most notable architectural treasure, St Peter's Church (see **frontispiece**), was built about 1160, and though partly rebuilt and altered many times, its Norman features are admirably displayed both inside and out. The brown sandstones and ironstone are from the local Northampton Sand, but the structure is richly decorated with limestone. The reassembled arch outside the west end is carved in cross-bedded limestone similar to the masonry (**11.11**); the columns of the corner buttresses are the same cream limestone, tending to soft weathering in patches, the matrix somewhat granular, with shell and some echinoderm debris. The piers inside mostly appear to be this limestone. It certainly comes from Northamptonshire, from the Great Oolite Group.

Of the Blisworth Limestone sources considered, the nearest geographically is Kingsthorpe (with which there was an ecclesiastical link), but there are no buildings of Blisworth Limestone there (the hilltop outcrop was dug in the nineteenth century, according to Sharp, for lime-burning and paving-stone); Raunds, though well situated along the Nene, and with a history of Saxon quarrying, has cross-bedded limestone with a different, rather heavier texture; a more likely source may have been Cosgrove, a place with ancient quarries, remnants of Norman architecture visible in its church, and a limestone quite similar to St. Peter's.

Helmdon Stone, from lower in the Great Oolite Group has also been considered; it was used for the Eleanor Cross in Northampton in the thirteenth century; the stone of the Cross (see **10.10**) is somewhat like the limestone at St. Peter's, but has more obvious fine layers. Some architectural pieces of a granular limestone from Northampton Castle and St James's Abbey may be comparable with St. Peter's. Detailed study of samples could be rewarding.

Above right & right: **11.11** The Norman Church of St Peter in Northampton (see **frontispiece**) was built *c*.1150, but the tower was rebuilt, further east, probably in the seventeenth century, and the limestone arch, of presumably a former west door, reset as decoration in the wall.

The limestone is cross-bedded, shelly and granular, with some echinoid spines. The buttress columns and most interior piers are similar stone, probably Blisworth Limestone Formation, but the source is not known.

THE 'MARBLES': RAUNDS AND STANWICK, DRAYTON AND ALWALTON

In the seventeenth and eighteenth centuries there was a vogue for a sparry, coarsely shelly 'raggstone' which could be polished like marble. Some came from the Lincolnshire Limestone at Weldon (Weldon Rag, see **1.7**), but also the 'very much esteem'd Rance Ragg', according to Morton, was used for the 'Chimney-Pieces and Window-Tables...of the Best Houses of this County'. The stone 'contains a beautiful variety of Shells, of both the Bivalve and Turbinated Sorts...the Bivalve Shells are the most numerous. They usually lye flatways in the Bed'. It was obtained from Blisworth Limestone quarries at Raunds, where the bed, believed by Thompson to have been about a metre thick, lay below three metres of limestone; he saw it to be much thinner (15

Above left: **11.12** The geologist Samuel Sharp towards the end of his life lived in Great Harrowden Hall (where he died in 1882). The fine house of Ketton Stone (*c.*1719) is now owned by the Wellingborough Golf Club. This fireplace may be Raunds 'Marble', a sparry limestone, full of shells (oysters and brachiopods) from the Blisworth Limestone, which was able to be polished.

Above right: **11.13** Drayton 'Marble' was dug in Drayton Park about 1700 for ornamental use in the House, including a chimney-piece in the King's Dining Room, and pedestals such as this one, full of oysters. Photographed with the permission of L. G. Stopford Sackville.

centimetres) at Stanwick. A few examples still exist: Great Harrowden Hall has a large chimney-piece (**11.12**), with oysters and brachiopods, that may be from Raunds, and there is a fireplace surround of Stanwick Marble in Stanwick Hall. A small monument of Raunds Marble is mounted in the parish church, and in Weekley church is one to Sir Edward Montagu (d. 1644). The marbles are buff to grey, and variable.

A similar shelly stone was dug in 1700 in the grounds of Drayton House, for the very fine chimney-piece in the King's Dining Room, and several ornamental pedestals (**11.13**). The polished stone is buff to dark olive-grey, with many oysters, seen sickle-shaped in side view, or as broader shells on the upper surface of the bed.

The most historically interesting and beautiful examples of 'marble' came not from Northamptonshire, or even the Soke, but were quarried from the Blisworth Limestone along the opposite (south) bank of the Nene at Alwalton, in Cambridgeshire, for the medieval effigies of the abbots in Peterborough Cathedral, carved in the twelfth and thirteenth centuries (**11.14**). Alwalton Marble (**11.15, 11.16**) is dark and crowded with oyster shells (unlike Purbeck 'Marble' which contains the roundish sections of freshwater snails). Shafts of the marble on the west front are paler, the oyster sections darker. The quarry was about seven kilometres by river from the cathedral.

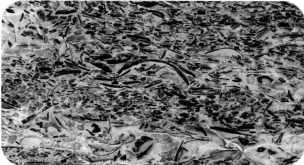

Right: **11.14** Alwalton Marble, quarried by the Nene, was carved and polished for a number of very fine tombs in Peterborough Cathedral, including this effigy of the Abbot Benedict (d.1193).

Top:: **11.15** Magnified view of Alwalton Marble, onto the bed of many oysters. (LEIUG23209).

Above: **11.16** Magnified side view of the bed of Alwalton Marble., packed with oysters. (LEIUG119550).

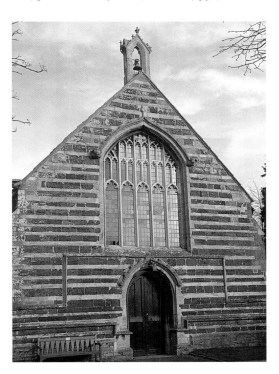

NORTHAMPTONSHIRE POLYCHROME

A very distinctive pattern of stonework encountered in many parts of the county was produced by the combination of two contrasting rock-types, usually Blisworth Limestone and brown Northampton Sand, and most commonly in stripes. It is a feature in places where these two rocks are found close together, the limestone outcrop lying just a few metres higher on the hillside. The effect is most striking where the Northampton Sand is a dark limonitic ironstone, and several such examples survive from the medieval period. They are common in parts of the Nene valley and the south-east; best known perhaps is the Bede House in Higham Ferrers, built by Archbishop Chichele in the fifteenth century (**11.17**) and to the south is St. Mary's Church, Rushden. In Wellingborough, the

11.17 The Bede House in Higham Ferrers churchyard was founded by Archbishop Chichele in 1422, and built of ironstone alternating with Blisworth Limestone, both obtainable locally. The dressings are of Stanion-type Lincolnshire Limestone.

Above: **11.18** This seventeenth-century cottage at Blisworth used to be familiar to travellers before the construction of the by-pass. There are several attractive examples of polychrome banding in the village. Ironstone (Northampton Sand) and Blisworth Limestone are from local outcrops; the windows are brown Northampton Sand from Duston.

Below: **11.19** The seventeenth-century Manor at Rothersthorpe is a handsome polychrome building of Blisworth Limestone and Northampton Sand, neither of them local. They would have been brought specially, the sandstone from the Duston area (possibly also the limestone), about 6 kilometres.

polychrome tradition established in medieval buildings was revived in the nineteenth century for the church of All Saints (see **4.7**).

Many limestone cottages in Blisworth are striped with dark brown ironstone; the purpose may have been decorative, but possibly also strengthening, ironstone being a tougher stone than the local limestone. Mullions were made of other Northampton Sand (probably from Duston) which was more easily shaped than the local ironstone (**11.18**). One of the best examples of two-tone banding is the seventeenth-century Manor at Rothersthorpe; however neither stone would have been obtainable in the village (which lies on Marlstone Rock), the Northampton Sand coming from Duston, the limestone possibly from quarries on Hopping Hill, (**11.19**).

An entirely different style of decorative stonework is seen at Stoke Park, Stoke Bruerne. Two pavilions are all that remain of the Palladian building of 1635 which has been attributed to Inigo Jones (**11.20**). The masonry is cream Blisworth Limestone, the pavilions ornamented by pilasters of contrasting brown ironstone from the Northampton Sand, both rocks being obtained within a few kilometres.

Other stone combinations have been employed in banding, some local and convenient, others brought for the purpose: the houses in Dallington (see **10.2**) are brown sandstone and a local limestone from below the Blisworth Limestone. The church tower at Irchester is handsomely striped, but in place of local Blisworth Limestone is a more sparry oolite, perhaps from the Cotswolds. The tower of Whiston Church (1534) is of oolitic Weldon freestone, with bands of brown sandstone, possibly from Duston. And Tresham's famous Triangular Lodge near Rushton (see **9.7**) has striped masonry of local ironstone and varied Lincolnshire Limestone.

CHAPTER TWELVE

History in Stone –
Brixworth Saxon Church

Many buildings hold a substantial record of their history, which can be recognised in a succession of architectural features, and visible adaptation of the masonry. The geological study of stonework can produce additional information for the architectural historian – especially where the building has a long history and the masonry is sufficiently varied.

12.1 The well-known Saxon Church of All Saints', Brixworth, from the south-west. The nave, (originally flanked with side-chapels that became aisles, now long gone), and the lower part of the tower, date from the eighth century. The stair turret and adjacent next stage of the tower were added about the eleventh century; the top of the tower being rebuilt and the spire added in the fourteenth century. The Saxon features were exposed and restored in the nineteenth century. The interior, also, is unique.

Above: **12.2** The small door on the south side of the tower led from a chamber of the former western narthex, long demolished. The arch, like the others of the eighth-century building, is made of narrow Roman bricks. The masonry includes an assortment of rocks not otherwise seen in Northamptonshire, such as various diorites and a large block of Triassic sandstone in the door jamb.

Above right: **12.3** The exotic masonry, seen here in the south wall of the tower, includes banded volcanic tuff, speckled diorites (one type is known as markfieldite), and dark metamorphic slate, similar to rocks occurring in Charnwood Forest and elsewhere in Leicestershire.

The Saxon Church of All Saints in Brixworth was built largely in the eighth century, on the site of a monastery founded in 680; and built of such an assortment of stone that the geologist W.J. Arkell called it 'a museum of rock types'. Brixworth is a large church by any standard (**12.1**), and when first built it was still larger, for the north and south aisles (each of which began as a row of separate chapels, or porticus) were demolished sometime in the Middle Ages. The nave, chancel and west tower are together 37 metres long, and at the east end is an apse, surrounded by an ambulatory below ground (no longer roofed); the tower originally had side-chambers forming a narthex, and a great west door, but this was altered when the later Saxon stair turret was built against it. The earlier Saxon arcades and clerestory were exposed and restored in the nineteenth century by the Vicar, The Reverend C.F. Watkins, by removing various later medieval windows; new windows inside the restored arches were surrounded by Harlestone sandstone, and neat Boughton Stone (light brown 'Pendle') used where necessary in the voussoirs. He also rebuilt the

polygonal apse, retaining recognisable Saxon features on the north side. His earlier work (1840) on the church's north wall included a plinth, built of shelly ironstone that probably came from the local Brixworth stone-pit south of the Workhouse.

In 1974 the Brixworth Archaeological Committee, under the chairmanship of Professor Rosemary Cramp, had already begun a survey of the building using the technique of photogrammetry, and now planned to improve the record with stone-by-stone scale drawings of the whole rubblestone fabric. Dr David Parsons from Leicester University organised the survey, which was carried out by teams of volunteers over several seasons, using scaffolding to cover successive elevations of the whole building (including interior walls when opportunity arose). Each stone was identified as the work progressed, and the geology coded on the drawings; there are more than 30 different types of stone, some of them being varieties of the local Northampton Sand, but including others not seen anywhere else in Northamptonshire. There are also quantities of Roman-style brick – in the walling and, conspicuously, fan-wise around the many arches, big and small.

The exotic stones can be studied easily from ground level; they occur in the masonry of the tower, nave and chancel (choir) up to a height of about 4 metres (**12.2**). Many of them are igneous rocks, visibly crystalline: various diorites, some with pink feldspars and dark greenish minerals, others reddish with white feldspars; and granite, which also contains quartz. There are striped grey rocks which are an indurated volcanic ash (tuff), and some dark grey, elongate blocks with the potential to split into slate (**12.3**). Along with these are large blocks of sandstone, some set in the sides of the arches,

12.4 The Roman masonry known as the Jewry Wall in the centre of Leicester is the remains of a bath complex. The rocks include diorites, volcanic tuff, granite, slate and Triassic sandstones from around Leicester; with a great deal of brick. Roman Leicester is considered to have been a source of materials for the builders of Brixworth Church in the eighth century.

12.5 This block of sandstone, with part of an inscription: 'ILIO', is evidence of re-used Roman material; it came from the foundations of the north side of the church during Michel Audouy's archaeological excavation in 1981. It is pinkish buff-coloured, probably Triassic sandstone.

including sugary white and various red sandstones; also several kinds of limestone.

The origin of this extraordinary assortment was discussed by visiting members of the Geologists' Association in 1921, who concluded that the rocks came from the local glacial deposits. But when the rocks of the church are compared with what is available in the boulder clay or gravels, they do not match at all, igneous rocks especially being very rare. Many of the rocks in the church are known to occur in Leicestershire, though not all in one place: some can be matched with the waterlain volcanic tuffs of Charnwood Forest, or the diorites (including 'markfieldite') that were intruded molten into the tuffs, and with the Swithland Slates in the same area; but others are like the diorites of Croft and Stoney Stanton; or like the granite of Mountsorrel, even including a distinctive altered rock (with mica and garnet) that is known to occur close to the Mountsorrel intrusion. There are not many of each, and they are unlikely to have been deliberately quarried at so many distant sources. The Roman bricks provide a clue; part of the remains of the Roman town still visible in the middle of Leicester (a site known as the Jewry Wall; 12.4) is built of assorted rocks, including diorites and granite, from various sources around Leicester, with a great deal of Roman brick, in courses in the masonry, and in the large arches which are similar in style to those at Brixworth. There is evidence that the Brixworth materials are re-claimed, some of the bricks having old (and not local) pebbly mortar adhering; and in the archaeological excavations of 1981 a sandstone block with part of a Latin inscription was recovered from the northern foundations (12.5); this certainly did not come from the 'Drift'! The exotic igneous

Opposite page: 12.6 Scale drawing of the west nave wall from the west side, with geological colour coding. (The central section is inside the tower.) Pink and purple are igneous and associated Leicestershire rocks; green, Triassic sandstones; yellow, limestones; shades of brown and orange, varieties of Northampton Sand; black, brick; light blue, Lincolnshire Limestone; deep blue, tufa. The significance attached to the distribution of the various stones is explained in the text.

From Sutherland and Parsons (1984), with permission from the British Archaeological Association.

12.7 The curved stair turret added in the eleventh century used Blisworth Limestone from elsewhere. The random burnt blocks indicate this was re-used stone.

carved oolite low in the nave west wall, similar to the Eagle (8th-9th century) in the south doorway, was probably inserted later, and there is a small window similar to Barnack Rag in the later turret. It seems that no building materials were supplied from Peterborough. Instead, access to material from Roman sites, which had become Crown property, suggests a Royal interest in the building of the great church at Brixworth in the eighth century, possibly by the powerful King Aethelbald of Mercia (716-757), a man known to have been influenced by the saints, Boniface and Guthlac; a reliquary at Brixworth was said to contain a relic of St. Boniface.

Petrological identification also yielded information about how the building developed. The survey effectively produced a geological map of the masonry (see 12.6 on page 115). The exotic assortment began with the foundations, including the porticus, and occupies the lower masonry of the nave and tower, but stops suddenly above the level of the west door, lowering to the east. Above that the stone is a mixed assortment of Northampton Sand. The reason for the break is not known; it may just mark the end of a building season, beginning a new one with different stone closer to hand, or there may have been a long gap in the project. The clerestory continues up with a mixture, and some batches, of Northampton Sand (ironstone, sandstone and calcareous types including 'Pendle') all occurring in the local geology. Along the line of the former aisle roofs the masonry is reddened by burning, especially noticeable on the north side, and the fire may have led to the eventual dismantling of the aisles; but significantly, many burnt stones are randomly distributed among unburnt stone through the Northampton Sand masonry, leading to the conclusion that much of it is re-used material, reclaimed from other fairly local earlier buildings.

In the west nave wall is a blocked arch directly above the door to the tower, presumably once allowing access between a nave gallery and the tower at this level. It was evidently blocked off during major alterations, when the stair turret was built, probably in the eleventh century. It is of interest to look at the exterior stonework of the turret in relation to the adjacent tower, and the stone 'map' of the west nave wall. The turret (12.7) is different from the earlier Saxon building, beginning with Blisworth Limestone, and no Leicestershire stone or Roman brick; but again there are random burnt blocks – this

rocks and sandstones appear to have come from Leicester. It is indeed surprising that the Saxons went so far, some 35 kilometres of hilly overland route. The limestones however are not from Leicester. They could have come from elsewhere in Northamptonshire, possibly other Roman ruins. Though the monastery was founded, we are told, as a daughter-house of Medeshamstede in 680, there is virtually no Barnack-type stone at Brixworth; a worn piece of

12.8 View of the turret and tower above about 4 metres; the corresponding masonry at each level indicates that they were built at the same time. Limestone was used first (above the pre-existing lower part of the tower), then joined by brown Northampton Sand, and then by tufa (a spongy-looking rock), seen as random blocks with Northampton Sand, and also alternating round the windows. The tufa was being used particularly for the stair vaulting inside the turret.

is reclaimed stone. The textures resemble limestone in southern Northamptonshire, and Roman Towcester is a possible source. The tower has this same limestone coming in directly over the exotic assemblage, indicating that the tower masonry was until then only that high (more like a porch, with side chambers) but presumably roofed to enclose the arch over the west nave door; or it may have been surmounted by a timber structure. The arch in the west nave wall is also blocked in the same limestone, so probably at the same time.

Higher up, brown Northampton Sand comes into the masonry of both turret and tower and gradually takes over from the limestone. At these levels, a new stone makes an appearance, tufa (12.8). This is not to be confused with tuff (the volcanic ash mentioned earlier). Is is a spongy-looking limestone which is formed by the evaporation of lime-bearing groundwater, precipitating calcium carbonate, for example where a seepage emerges. Deposits are not uncommon in Northamptonshire, and one occurs at the head of the Nene tributary near Haselbech. Tufa was especially chosen for vaulting, as it is lightweight yet strong, and at Brixworth it was used to make the spiral staircase in the turret. As the turret went up, blocks of tufa were also used in the exterior masonry here and in the tower beside it, and can be seen lining some of the windows, the blocks alternating with Northampton Sand. Notice too (see **12.6**) that tufa also comes into heads of the triple arch created above the blocked earlier arch (with the Roman brick re-used yet again), and as packing in old beam-holes.

Tufa can also be seen in the Saxon remnants of the apse (**12.9**) (again alternating with blocks of Northampton Sand), enabling this to be related to the same construction phase). It is not certain whether this was refurbishment of an already existing apse and ambulatory. The latter is much covered by rendering, but there appear to be one or two stones of the exotic assemblage and certainly Roman brick springing to the former vaulting, suggesting that the ambulatory was an early feature, and the apse, with its tufa, a late-Saxon rebuild. The access to the ambulatory was probably improved at that time; the large blocks of the steps are similar to the foundations of the turret. The eleventh-century refurbishment evidently involved the whole building, re-using putlogs for scaffolding, which were finally blocked with left-over tufa.

This brief account is a summary of the church's early history, insofar as it can be deduced from a study of the masonry; it is but an introduction to the most remarkable historic building in the county.

12.9 The Saxon remnants of the apse (most was rebuilt in the nineteenth century), with tufa and Northampton Sand around the window and forming a pilaster.

Conclusion

Although many stone buildings are known to have been demolished, fortunately a great number remain. Some of the larger country houses are still the homes of the landed aristocracy, and some, such as Boughton House and Rockingham Castle, are in season open to the public. Lamport Hall is now run by a Trust, and is also accessible as advertised. Kirby Hall is in the care of English Heritage. Also in this county of stone villages, there are many which still have a Hall; some are privately owned, some restored by new owners (including Canons Ashby House, restored by The National Trust); some have become schools, and others, from Dallington to Dingley, have been converted to apartments, giving the elegant architecture a new lease of life. Almost every village has a historic stone church; and cottages which once may have been the 'meaner houses' noted by John Morton in 1712, are now the desirable listed buildings of the twenty-first century.

Whilst we have come to appreciate and value our stone buildings, the same cannot be said of the quarries that once supplied the stone. Some were lost in the later quarrying for ironstone over much of the county – most obviously Weldon near Corby, Finedon, Desborough, and Brixworth – and the extensive quarries were mostly back-filled and returned to agriculture or other uses, a few becoming nature reserves. Others have clearly been overridden by the spread of development, the expansion of Northampton, for instance, successively covering all its former stone pits: first, in the town, then those in the 'Fields', those around the racecourse, then those of Kingsthorpe, and within the last twenty years, the great quarry at New Duston. Out in the villages, too, old historic quarries such as Eydon and Cosgrove have been filled in. People in tidy villages do not want quarries.

The few remaining quarries mostly supply crushed stone for aggregate, with some local building stone. When stone is needed for renovation of any building, it now usually comes from outside the county. In the case of Marlstone Rock, sources in Oxfordshire are a close match, if the quality is good. The paucity of stone for renovating Northampton Sand buildings is less satisfactory especially as there are many different varieties to consider in selecting the most appropriate. Use is made of Hornton Stone, ginger Carstone (sometimes both, as in All Hallows', Wellingborough), and regrettably much recourse to yellow Guiting limestone which is nothing like the brown sandstones (it is however quite similar to the golden 'Pendle'). Fortunately, Harlestone quarry can supply brown sandstone, but varieties of local ironstone are more difficult to obtain. Collyweston stone slate is also in short supply, and fewer slaters are working it. Varieties of good Lincolnshire Limestones are available from beyond the county, the porous oolite from Ketton being perhaps nearest match to Weldon and King's Cliffe. Pury End is the only supplier of Blisworth Limestone, since Churchfield quarry near Oundle closed.

The fortunes of the quarrying industry wax and wane, for various reasons. But with the many historic stone buildings needing a supply of appropriate stone for restoration, perhaps the county's landowners will be encouraged to keep existing quarries open, and to consider finding new sources.

Bibliography

GEOLOGICAL MAPS
British Geological Survey (www.bgs.ac.uk)
1:50,000 or 1inch to a mile
Stamford Sheet 157
Market Harborough Sheet 170; Kettering Sheet 171
Warwick Sheet 184; Northampton Sheet 185;
Wellingborough Sheet 186
Banbury Sheet 201; Towcester Sheet 202; Bedford
(old series)
Chipping Norton Sheet 218; Buckingham Sheet 219
1:250,000: East Midlands (without drift)

Arkell, W.J., 1947. *Oxford Stone.*

Audouy. M., 1984, Excavations at the Church of All Saints', Brixworth, Northamptonshire. *Journal of the British Archaeological Association, 137, 1-44*

Aveline, W.T. and Trench, R., 1860. Geology of parts of Northamptonshire. *Memoir of the Geological Survey.*

Baker, G.T., 1823-30, and 1836-41. *The history and antiquities of the county of Northampton.*

Best, J., 1978. *Using the environment, No.6 - (a) Quarries, Weldon and Ketton.* Nene College, Northampton.

Binney, M., 1971. Eydon Hall. *Country Life,* p.128.

Bradshaw, M.J., 1978. 'A facies analysis of the Bathonian of eastern England.' Unpublished D.Phil. thesis, University of Oxford, 6 vols.

Bradshaw, M.J. and Cripps, D.W., in: Cope, J.C.W., Ingham, J.K. & Rawson, P.F. (eds), 1992. Atlas of Palaeogeography and Lithofacies. *Geological Society, London, Memoir, 13.*

Bridges, J., 1791 (Ed. P.Whalley). *History and antiquities of the county of Northampton.*

Briggs, N., 1991. John Johnson 1732-1814. *Essex Record Office Publication No.112.*

British Museum (Natural History), 1983. *British Mesozoic Fossils.*

Cadman, G., 1990. Recent excavations on Saxon and medieval quarries in Raunds, Northamptonshire. In: Parsons, D., *q.v.* , 186-206.

Clifton-Taylor, A., 1972. *The pattern of English building.* Faber paperback edn.

Clifton-Taylor, A. & Ireson, A.S., 1983. *English stone building.* Victor Gollancz.

Colvin, H.M., 1978. *A biographical dictionary of British architects.*

Cox, the Rev. J.C. & Serjeantson, the Rev. R.M., 1897. *History of the Church of the Holy Sepulchre, Northampton.*

Dryden, A., 1903. *Memorials of old Northamptonshire.*

Edmonds, E.A., Poole, E.G. and Wilson, V., 1965. The geology around Banbury and Edge Hill. *Memoir of the Geological Survey.*

Fenton, J.P.G., Riding, J.B., and Wyatt, R.J., 1994. Palynostratigraphy of the Middle Jurassic 'White Sands' of central England. *Proceedings of the Geologists' Association, 105, 225-230.*

Field, E.E., 1930. Abington village. [with map of 1840]. *Journal of the Northants. Natural History Society, 25, 174-177.*

Field, E.E. and Thompson, B., 1930. The parish of Abington. [with map 1671]. *Journal of the Northants. Natural History Society, 25, 146-154.*

Gotch, J.A., 1883. *A complete account of the buildings erected by Sir Thomas Tresham.*

Gover, J.E.B, Mawer, A., and Stenton, F.M., 1933. *The place-names of Northamptonshire.*

Greenall, R.L., 1979. *A history of Northamptonshire and the Soke of Peterborough.* Phillimore.

Hains, B.A. and Horton, A., 1969. *British Regional Geology: Central England.* HMSO.

Heward, J. and Taylor, R., 1996. *The Country Houses of Northamptonshire.* Royal Commission on the Historical Monuments of England.

Hewit, R., 1692. *Survey and veiw of the Mannor of Duston.*

Hill, C., Millett, M., and Blagg, T., 1980. The Roman riverside wall and monumental arch in London. *London and Middlesex Archaeological Society, Special Paper, 3.*

Hollingworth, S.E. and Taylor, J.H., 1946. An outline of the geology of the Kettering district. *Proceedings of the Geologists' Association, 57, 204-233.*

Hollingworth, S.E. and Taylor, J.H., 1951. The Northampton Sand Ironstone: stratigraphy, structure and reserves. *Memoir of the Geological Survey.*

Horton-Smith, L.G.H., 1943. The later Lumleys of Harleston. *Journal of the Northants. Natural History Society, 30, 74-108.*

Howarth, M.K., 1978. The stratigraphy and ammonite fauna of the Upper Lias of Northamptonshire. *Bulletin of the British Museum (Nat. Hist.), 29, 235-288.*

Howarth, M.K., 1992. Hildoceratidae in the Lower Jurassic of Britain. Part I. *Palaeontographical Society Monograph.*

Hudson, J.D. & Sutherland, D.S., 1990. The geological description and identification of building stones: examples from Northamptonshire. In: Parsons, D. (Ed.). *q.v.,* 16-32.

Hussey, C., 1951. Ashby St. Ledgers, Northamptonshire. *Country Life,* 274-277, 348-351, 420-423, 496-499.

Isham, G., 1966. Sir Thomas Tresham and his buildings.

Reports and Papers of Northamptonshire Antiquarian Society, Part II.

Jenkins, H.J.K., 1993. Medieval barge traffic and the building of Peterborough Cathedral. *Northants Past & Present*, 8, No.4, 255-261.

Jope, E.M., 1964. The Saxon building-stone industry in southern and Midland England. *Medieval Archaeology*, 8, 91-118.

Judd, J.W.,1875. The geology of Rutland etc. *Memoir of the Geological Survey.*

Kerr, W.J.B., 1925. *Higham Ferrers and its Ducal and Royal Castle and Park.*

Maguire, H.P., 1970. A twelfth century workshop in Northampton. *Gesta*, 9, 11-25.

Morton, J., 1712. *The natural history of Northamptonshire.* London.

NRO (Northamptonshire Record Office): Byfield: 1779 Map 3495; Duston: 1722 Map 6013; Finedon: 1805 Map 625; Kingsthorpe: 1767 Map 2845; Stanion: 1730, Map 4340

Oswald, A., 1951.Winslow Hall, Buckinghamshire. *Country Life*, 572-576.

Palmer, J. and Palmer, M., 1972. *A history of Wellingborough.*

Parry, E.G., 1987. Helmdon Stone. *Northants Past & Present*, 7, No.4, 258-269.

Parsons, D.(Ed.), 1990. *Stone: Quarrying and building in England AD 43-1525.* Phillimore and R.A.I.

Pevsner, N. & Cherry, B., 1973. *The buildings of England. Northamptonshire.* Penguin.

Poole, E.G., Williams, B.J., and Hains, B.A., 1968. Geology of the country around Market Harborough. *Memoir of the Geological Survey.*

Purcell, D., 1967. *Cambridge stone.* Faber & Faber.

RCHME, 1984 *An inventory of architectural monuments in north Northamptonshire.* Royal Commission on Historical Monuments of England. HMSO, London.

Richardson, L., 1925. Certain Jurassic (Aalenian) strata of the Duston area, Northamptonshire. *Proceedings of the Cotteswold Nat. Field. Club*, 22, 137-152.

Salzman, L.F., 1952. *Building in England down to 1540.*

Scott, S., 1995. *The follies of Boughton Park.*

Serjeantson, the Rev. R.M., 1904. *History of the Church of St. Peter, Northampton.*

Sharp, S., 1870. The oolites of Northamptonshire. *Quarterly Journal of the Geological Society of London*, 26, 354-393.

Sharp, S., 1873. The oolites of Northamptonshire. Part II. *Quarterly Journal of the Geological Society of London*, 29, 225-302.

Steane, J.M., 1967. Building materials used in Northamptonshire and the area around. *Northants Past & Present*, 4, 71-83.

Sutherland, D.S., 1990. Burnt stone in a Saxon church and its implications. In: Parsons, D. (Ed.), *q.v.*, 102-113.

Sutherland, D.S. & Parsons, D., 1984 (1985). The petrological contribution to the survey of All Saints'

Church, Brixworth, Northamptonshire: an interim study. *Journal of the British Archaeological Association*, 137, 45-64.

Sylvester-Bradley, P.C. and Ford, T.D., 1968. *Geology of the East Midlands.* Leicester University Press.

Taylor, J.H., 1963. Geology of the area around Kettering, Corby and Oundle. *Memoir of the Geological Survey.*

Thompson, B., 1888. *The Middle Lias of Northamptonshire.*

Thompson, B., 1884-1888. The Upper Lias of Northamptonshire. Parts I-VI. *Journal of the Northants. Natural History Society.*

Thompson, B., 1891. The oolitic rocks at Stowe-Nine-Churches. *Journal of the Northants. Natural History Society*, 6, 295-310.

Thompson, B., 1921. Excursion to Northamptonshire. *Proceedings of the Geologists' Association*, 32, 219-226.

Thompson, B., 1927. *Lime resources of Northamptonshire.* Northamtonshire County Council.

Thompson, B., 1928. *The Northampton Sand of Northamptonshire.* (Reprinted papers in the *Journal of the Northants. Natural History Society* from 1921) Dulau.

Thompson, B., 1930. The Upper Estuarine Series of Northamptonshire and northern Oxfordshire. *Quarterly Journal of the Geological Society*, 86, 430-462.

Till, E., 1998. Fact and conjecture - the building of Burghley House: 1555-1587. *Northants. Past & Present*, 9, No.4, 323-332.

Tonks, E., 1988-1992. *The ironstone quarries of the Midlands*, Parts II-VI

Torrens, H.S., 1967. The Great Oolite Limestone of the Midlands. *Transactions of the Leicester Literary & Philosophical Society*, 61, 65-90.

VCH: *The Victoria history of the county of Northampton*, Vol I-V.

Vol. II, 1906: Thompson, B., Quarries and Mines (Technical), 298-307. Vellacott, C.H., Quarries (Historical),293-298.

Waddy, F.F., 1974. *A history of Northampton General Hospital.*

Wake, J. and Webster, D.C., 1971. The letters of Daniel Eaton to the 3rd Earl of Cardigan 1725-1732. *Northants Record Society XXIV.*

Williams, J.H., 1971. Roman building materials in south-east England. *Britannia*, 2, 166-195.

Winchester, S., 2001. *The map that changed the world. The tale of William Smith and the birth of a science.* Viking.

Wood-Jones, R., 1963. *Traditional domestic architecture of the Banbury region.* Manchester University Press.

Wormald, P., 1991. The age of Bede and Aethelbald. In: Campbell, J. (ed.). *The Anglo-Saxons.*, 70-100. Penguin edition.

Glossary

Ammonite: extinct cephalopod mollusc having chambered shell, commonly in a planar coil

Ammonite Zone: time interval defined by a particular ammonite species

Ashlar: squared masonry with a smooth surface and close joints with little mortar

Bed: a thick or thin layer of soft or hard sedimentary rock, separated from beds above or below by a recognisable break known as a **bedding plane**

Belemnite: extinct squid-like cephalopod mollusc with internal solid calcite skeleton; elongate, bullet-shaped guard is the usual part preserved

Bivalve: aquatic mollusc having calcareous shell of two valves (e.g. oyster, mussel)

Blue-hearted: a blue-grey limestone weathering to cream or buff, by the oxidation of particles of pyrite (iron sulphide) dispersed in the rock

Box-stone: concentrically layered structure of hard limonite, in weathered ironstone

Boulder clay: see **Till**

Brachiopod: lamp-shell; marine invertebrate having two valves; e.g. **Terebratulids** and **Rhynchonellids**

Brackish: variably salt water, less salty than sea-water

Calcareous: containing calcium carbonate, e.g. as calcite cement, or any shelly particles

Calcite: common crystalline form of calcium carbonate

Carbonaceous: term applied to sediment containing carbon, usually organically derived, such as plant remains

Cement: the mineral filling the pore-space in a sedimentary rock

Channel: the course of a relatively narrow zone of moving water, eroding the underlying rock

Crinoid: an echinoderm, known as a sea-lily; most having a stem attached to sea-floor

Cross-bedding: fine layering in a sediment, inclined at an angle to the main bedding; produced by current action (formerly **Current-** or **False-bedding**)

Diorite: a coarsely crystalline igneous rock, darker and with less quartz than granite

Drift: superficial sediments deposited over Solid bedrock

Echinoderm: member of marine invertebrate phylum that includes sea urchins and sea-lilies

Erosion: the wearing away of the earth's surface, by wind, water, or ice

Feldspar: the most common mineral of igneous rocks, as white or pink crystals

Ferruginous: containing iron; a sedimentary rock containing iron minerals

Fissile: having a tendency to split in thin layers

Fossils: the remains of animals or plants sometimes preserved in sediments

Fossiliferous: describes sedimentary rock containing fossils

Freestone: quarry-stone able to be cut freely in any direction; often obtainable in large blocks

Glacial: associated with ice; deposits carried by or derived from ice sheets or glaciers

Goethite: mineral form of the hydrous iron oxide, limonite

Grain-size: sedimentary particles are classed as mud<0.063mm <sand<2.0mm< gravel. Igneous rocks are coarse->5mm>medium->1mm>fine-grained

Granite: coarse-grained crystalline igneous rock containing feldspar and quartz

Igneous: rocks that were once molten

Intrusive: molten rock that has pushed its way up into solid rocks, and cooled as an **Intrusion**

Ironstone: a sedimentary rock rich in iron minerals; or workable deposit yielding over 20 per cent of metallic iron

Joint: a fracture in rock, especially in a set of planes having common direction

Jurassic: a period of time between about 210 and 140 million years ago

Laminated: thinly layered sedimentary rock

Limestone: a sedimentary rock composed of calcium carbonate

Limonite: hydrous oxide of iron, ochreous or brown, formed by weathering of iron minerals (pronounced as in 'lime', but there is no lime in it)

Ma: short for 'millions of years ago'

Marl: a soft rock composed of a mixture of clay and fine calcium carbonate

Marble: a recrystallised, metamorphosed limestone; or sparry limestone able to be polished

Metamorphic: rock that has been subjected to heat or pressure, producing new crystals or structure

Mica: a silicate mineral occurring as thin flat flakes

Micaceous: a rock containing mica

Micrite: very fine-grained calcium carbonate mud

Mineral: naturally occurring element or compound having consistent atomic structure

Mudstone: fine-grained (mud size) sedimentary rock e.g. clay

Nodule: a rounded lump formed by chemical concretion

Oolite: a sedimentary rock consisting largely of ooliths, usually calcium carbonate

Oolith: a spherical grain in a sediment, having concentric and radial structure, formed by precipitation, chemically or via biological agency

Oolitic: a sedimentary rock containing ooliths; applicable to limestone or ironstone

Outlier: a remnant of younger rock in area of older strata, for example isolated by erosion

Pendle: quarryman's term for limestone that splits into flags; used for certain sandy limestones in the Northampton Sand Formation

Pyrite: iron sulphide ('fool's gold')

Quartz: common crystalline form of silica (silicon dioxide)

Rag: hard stone suitable for rough walling; also a ossiliferous sparry limestone

Rhynchonella: Brachiopod with calcareous, ribbed shell

Rock: an aggregate of mineral material, which can be induratesd or loose particles

Rock-faced masonry: blocks tooled to a rough front surface, the sides flat, close-spaced

Rootlet marks: vertical carbonaceous traces of plants in sedimentary rock

Rubble, Rubblestone: walling of hammer-dressed, or only roughly shaped blocks, generally with obvious mortar

Sandstone: aggregate of sand grains (commonly quartz), held in mineral cement

Sedimentary rock: formed by accumulation of sediment, deposited on land or in water

Shelly: containing shell and/or calcareous fossil fragments

Siderite: the mineral iron carbonate (pronounced as in 'cider')

Silica: silicon dioxide, e.g. the mineral quartz

Slate: fissile, usually metamorphic rock, formed by pressure; **stone slate** is naturally fissile sedimentary rock

Spar: crystalline calcite

Superficial ooliths: small fragments of shell coated with layer of calcium carbonate

Terebratula: Brachiopod with calcareous smooth shell

Tertiary: period of time between 65 and 2 million years ago

Thin section: a thin slice of rock, ground further to 30 microns, which can be examined by transmitted light under a microscope

Till: mixture of clay, cobbles and boulders (boulder clay), deposited by melting ice

Triassic: period of geological time between about 250 and 210 million years ago

Tufa: sedimentary deposit of calcium carbonate formed by evaporation, e.g. around springs

Tuff: volcanic ash deposit, either loose particles (<2 mm) or indurated

Unconformable: strata which rest on earlier ones after a significant gap in sedimentation

Unconformity: a gap in sedimentation, a period of erosion; a break in sequence of strata

Water-table: (architectural) see **Weathering**

Weathering: alteration of rock exposed at the earth's surface: physical break-up by frost, and chemical oxidation and hydration forming new minerals; architecturally, a sloping coping stone or **water-table** (e.g. on a buttress) to shed rainwater

Index